Creating Abundance

Creating Abundance

America's Least-Cost Energy Strategy

Roger W. Sant
Dennis W. Bakke
Roger F. Naill

James Bishop, Jr., Editor

McGraw-Hill Book Company

New York   St. Louis   San Francisco   Auckland   Bogotá   Hamburg
Johannesburg   London   Madrid   Mexico   Montreal   New Delhi
Panama   Paris   São Paulo   Singapore   Sydney   Tokyo   Toronto

Library of Congress Cataloging in Publication Data
Sant, Roger W.
    Creating abundance.

    Includes index.
    1.Power resources—United States.   2.Energy policy—
United States.   I.Bakke, Dennis.   II.Naill, Roger F.
III. Title.
TJ163.25.U6S26   1984   333.79′0973   83-24853
ISBN 0-07-041518-8 PBK.

1234567890   DOCDOC   89876543

ISBN 0-07-041518-8 PBK.

The editors for this book were Joan Zseleczky and Esther Gelatt,
the designer was Mark E. Safran, and the production supervisor
was Thomas G. Kowalczyk. It was set in Baskerville by Byrd Data
Imaging Group.

Printed and bound by R. R. Donnelley & Sons, Inc.

Contents

Preface

This book is less a call to action than it is a documentation that the once awesome energy battle is being won.

After a decade of exhausting debate about energy problems, a time remembered for deep divisions of opinion as well as extravagant claims from all quarters, the nation is proceeding rapidly down a different path. High expectations about government effectiveness in the energy arena have produced great disappointments, yet our energy problem is being solved. The new energy path is being formed by individual decisions, not by governments. The unexpected bonus of this new path is the growing acknowledgment of how effective a wide range of individual responses has become.

After sifting through the fallout of years of crisis-inspired energy remedies, two of us created the Mellon Institute's Energy Productivity Center in 1977 to develop and test a fundamentally different hypothesis. It called for treating energy like any other commodity whose price and supply are uncertain, rather than continuing to deal with it under the aegis of a complex web of government controls and subsidies.

Energy, after all, is nothing more than a class of commodities produced, traded, and consumed predominantly in the private sector—similar to, for example, primary foodstuffs, metals, or forest products that are also incorporated into finished goods and used by consumers. We reasoned that, as with all commodities, relative costs rather than social goals determined the processes by which these goods are used and produced.

In constructing this hypothesis, the social dimensions of the global energy situation were not ignored. But we argued that the nation's energy system was altogether different from the national defense establishment or the space program, for example, where government has historically been assigned operating responsibility for many of the technical systems. Unlike the space or defense programs, energy is produced and consumed largely in the private sector.

To be sure, there were obvious dangers in this new approach. If various fuel sources were treated as commodities rather than scarce national resources, individuals and companies would be expected to act primarily in their own interest as they produced and consumed petroleum, natural gas, electricity, and other fuels. As a result, the amount and the mix of energy produced would adjust so that only the most competitive energy forms would survive. If so, what would happen to low-income consumers, the environment, or national security? If our hypothesis were to become reality, was it possible that matters would be made worse in those areas?

In order to assess the consequences of adopting the commodity approach for energy, in the fall of 1979 we constructed what we called the Least-Cost Energy Strategy[1]—an analysis of how the energy system would function and evolve if it were to be guided by economic forces alone. As a basis for this strategy, all regulations affecting energy choices were assumed to have been jettisoned. It was also assumed that consumers would pick various energy systems for their homes, factories, and vehicles according to their own direct economic self-interest; that is, they would try to minimize their costs. The results of the analysis formed the basis for a new optimism. Instead of the Least-Cost approach being a recipe for social disaster, we found that the reverse was true.

What was unique about the calculations for the Least-Cost Energy Strategy? The main departure from more traditional approaches was the decision to regard energy only as a means of providing the services that a modern economy requires—such as mobility, comfort, or industrial heat. We reasoned that people simply want comfort in their homes at the least cost to them; if costs are equal, few will put up a fight over whether they prefer electricity to natural gas. Similarly, industrial users need services like steam for chemical or refinery processes or shaft power to run machines and equipment. They care little whether they use

---

[1] Roger W. Sant, *The Least-Cost Energy Strategy*, Carnegie-Mellon University Press, Pittsburgh, 1979.

natural gas, oil, or coal, as long as they feel they are reliably obtaining the services they need at the lowest possible cost.

When U.S. energy services were analyzed that way, it became obvious that fuel or electricity was only a part of what was required to provide a useful service. For example, creating comfortable rooms requires not only fuel oil or natural gas but a furnace, thermostat, duct work, insulation, and numerous other support items. It became clear that what is really important is putting together all the components in such a way as to minimize the total cost of delivering energy services.

After establishing the energy services context, we calculated the combination of fuels and energy-using devices (such as more energy-efficient furnaces) that could have met the actual demand for energy services in 1978 at the lowest possible cost. The results—though only hypothetical, since we can never relive history—stood as a reasonable guide to what would be possible if simple economic self-interest were the principal motivating force behind our energy technology choices. The bottom line was that consumers were spending some $44 billion per year too much for their 1978 energy service costs.

That exercise was then updated so that, instead of assessing just the conditions in 1978, the Least-Cost fuel mix and investment patterns were projected through the year 2000.[2] Those projections were not presented as a forecast of what would happen but were intended to show what would be possible if available energy technology were to be applied to our energy future purely on the basis of economic self-interest. From the results of this work we learned that a host of our most common beliefs about energy were myths. One after another, the "truths" tumbled and fell apart under closer scrutiny—from the belief that "energy is scarce" to the conviction that "natural gas, when deregulated, will rise to the price of oil."

In the summer of 1982, almost a year after completing the Energy Productivity Center project, Dr. Roger Naill joined us as a coauthor. He and his staff reran the center's models, using a more advanced approach,[3] and found these same directions so strongly indicated that we label his update as a forecast in Chapter 7.

If this book had been written a few years ago, the overriding theme would have been that energy problems can be solved, the job can be done. Today, the theme is that the job is being accomplished right before our eyes. The nation is winning the battle against energy

[2] Roger W. Sant et al., *Eight Great Energy Myths*, Carnegie-Mellon University Press, Pittsburgh, 1981.

[3] Details of the 1982 analysis can be found in *The Least-Cost Update*, Applied Energy Services, Inc., 1925 N. Lynn Street, Suite 1200, Arlington, VA 22209.

problems. For in terms of energy productivity, the nation is doing better—better, in fact, than the original Least-Cost projections suggested. And from the projections we have now made, there is every indication that these winning ways will continue. Therefore this book is less a call to action than a documentation that the once awesome energy battle is being won. It is our way of describing the impressive accomplishments of Americans in taming the energy problem. It is a means of identifying some of the untapped opportunities that still exist. And it is our way of etching the lessons we have learned on the canvas of public awareness, to be used in the future when problems with similar characteristics arise.

Roger W. Sant
Dennis W. Bakke
Roger F. Naill

Acknowledgments

In the course of doing the research for and writing this book, we were assisted and supported by hundreds of people. Some of their contributions came from only brief contact, but other contributors dedicated huge amounts of time far beyond what any of us could have expected. Our deep appreciation extends to all of these people, but we would expressly like to mention some by name.

We start with Carnegie-Mellon University (CMU) and its contract research subsidiary, the Mellon Institute. It was they who sponsored the Energy Productivity Center (EPC). Dick Cyert, the president of CMU, gave us unfailing support throughout the 4 years of the center's existence. Without his confidence, we never would have completed the productivity project or this book. Ted Hermann, then president of the Mellon Institute, had the original idea for the center and did everything conceivable to ensure that it made a contribution to the energy dialog. His assistance in formulating the purpose of the center, obtaining initial funding, and providing administrative help was especially invaluable. Art Rolander, a consultant to the Institute and currently a member of the board of directors of AES, also played a key role in the start-up and ongoing activities of the EPC.

The center was generously supported by grants and contracts from the American Gas Association, the American Paper Institute, American Telephone & Telegraph Company, Atlantic Richfield Foundation, Department of Energy, Dravo Corporation, Edison Electric Institute, Electric Power Research Institute, Gas Research Institute, Gulf Oil Foundation, Honeywell Incorporated, International Business Machines Corporation, Koppers Company Foundation, Motor Vehicle Manufacturers Association, National Institute of Building Sciences, Ralston Purina Company, Rockwell International Corporation Trust, Sun Company, the Dow Chemical Company, 3M, and Westinghouse Electric Corporation. To all of these organizations we express our deep gratitude for allowing us the experience of exploring, dissecting, and evaluating the energy issue.

At the EPC itself, we were blessed with an extraordinary advisory board. Bill Kieschnick, now chief executive officer (CEO) of Arco, was chairman. We are still amazed at how, in addition to all his extensive responsibilities, he could donate the time to give us such valuable advice and counsel. Each of the other board members—Al Alm, now deputy administrator of the Environmental Protection Agency; Susan Lewis Borden, who was then in charge of energy programs at the Aspen Institute; Ted Burtis, the CEO of the Sun Company; George Hatsopoulos, CEO of Thermo Electron Corporation; Henry Linden, president of the Gas Research Institute; Russell Train, former administrator of the Environmental Protection Agency and now president of the World Wildlife Fund; Dana Meadows, coauthor of *Limits to Growth* and associate professor at Dartmouth College; Glenn Watts, president of the Communication Workers of America; Lee White, former chairman of the Federal Power Commission; Frank Potter, staff director of the House Committee on Commerce and Energy; and Jim Fletcher, former administrator of NASA and professor at the University of Pittsburgh—gave liberally of their advice, time, and personal support. They kept us constantly assessing and reassessing our work products.

The staff of the center, whose names should all appear on the cover of this book, were really the ones who made its publishing possible. We especially thank Steven Carhart, whose creativity and skill provided the basic models of the energy-consuming sectors that are still in use today and formed the basis for the analysis in the book. He and Shirish Mulherkar developed from scratch the buildings model and adapted the industrial model, developed superbly by Bob Reid and Energy and Environmental Analysis, Inc., and the transportation model, developed by Jack Faucett Associates.

Dick Shackson and Jim Leach did all the work relating to the transportation sector, providing a constant source of ideas about the future of automobiles and other transportation modes. Chapter 3 was almost entirely written by Jim Leach based upon this work.

Sandra Rennie headed the research in the buildings sector and, along with Alton Penz, is responsible for developing much of the content of Chapter 4. Their constant search for new means of delivering energy services in that sector significantly pushed along the progress of the nation.

Alan Hoffman took the existing industrial model and, supported by a contract from the Department of Energy, led the effort to make important improvements to that tool that are currently being utilized by ourselves and others. He and Debbie Long also provided much of the ancedotal material in Chapter 5.

The remainder of the EPC staff was equally superb: Althea Armstrong, Laureen Bakri, Walter Brooks, Yvonne Irvine, Leith Mann, Anjuna Mathi, Robin Miller, Barbara Payne, Carolyn Weisner, Susan Wrisley, and Pat Young.

The people at AES considerably expanded the work of the center from October 1981 onward. Jay Geinzer, Raffy King, Jay Meek, Sheryl Sturges, and Francis Wood did the analysis for the Least-Cost update that provides the primary data used in Chapters 7 and 9. Jay Geinzer also did the graphic work for the figures in this book. Corinne Langendoen and Susan Wrisley were incredibly patient and persistent in typing numerous drafts of the manuscript.

Anjuna Mathi took on the job of managing the project from beginning to end—making the authors' job much easier. When we got distracted, it was Anjuna who would bring us back to our purpose. Without her enthusiasm, we might still be working on the manuscript.

Several people reviewed drafts of the manuscript and provided very helpful suggestions and counterarguments: Amory and Hunter Lovins (whose proclivity continues to amaze us), Lester Brown, Frank Potter, Genia Potter, Al Alm, and Jack Gibbons. Nevertheless, none of these people bear any responsibility for the content of the book.

Finally, but most important, we express our loving appreciation to our wives—Vicki, Eileen, and Carol—and our children—Shari, Mike, Ali, Lex, Brett, Scott, Dennis, Sara, and Megan—for being so patient with us as we took time away from them to complete the manuscript. Their encouragement, support, and editorial suggestions made the process ever so much more fun.

A Note to the Reader

Most people are accustomed to measuring energy usage in terms of the physical quantities of the energy form used: tons of coal, barrels of oil, cubic feet of natural gas, or kilowatt-hours of electricity. To discuss the overall energy system, as this book does, it is helpful to convert these various units of measurement to a common unit of convenient size. One property that all fuels have in common is their *heat value*; in the English system of measurement, the basic unit is the Btu. A "Btu" is a British thermal unit, defined as the amount of heat needed to raise the temperature of 1 pound of water 1°F.

The following table gives some useful conversion factors:

Oil	5,800,000 Btu per barrel
Natural gas	1030 Btu per standard cubic foot
Coal	22,500,000 Btu per ton*
Electricity	3412 Btu per kilowatt-hour

* For medium heating, value coal at 11,250 Btu per pound.

Note that because of the inherent conversion losses in generating electricity from conventional heat sources, about 3 to 3.3 Btu of energy from coal, oil, gas, or uranium are required to produce 1 Btu of electrical output.

Since the Btu is a small unit compared to the amounts of energy currently used by the United States each year, it is more convenient to use the "quad"—1 quadrillion Btu, or 10^{15} Btu—when discussing the energy system.

1

Turning the Corner

The public has made up its mind that energy costs are out of proportion to other things they want out of life.

Fuel Oil Dealer

Know the power of self-interest in human society without giving it moral justification . . . beguile, deflect, harness and restrain self-interest, individual and collective, for the sake of the community.

Reinhold Niebuhr

From Maine's coastal villages westward to southern California and in many small towns and cities in between, Americans by the millions are grappling with energy problems and experimenting with solutions; fuel switching, home insulation, cogeneration, wood stoves, passive solar, high-mileage automobiles, efficient appliances, and van pooling are but a few of the proven ways to reduce energy costs. This new consciousness, fired by a combination of higher energy prices, the bitter memory of gasoline lines, and perhaps a certain amount of patriotism, has created a new national but highly diversified movement, one that has gone largely unrecognized in energy reporting. As one fuel oil dealer put it, "The public has made up its mind that energy costs are out of proportion to other things they want out of life." Because of their actions, the energy problem is taking a positive turn, a feat that was once considered impossible. Collectively, these efforts have worked to reduce energy costs for all consumers from what they otherwise would have been. As

1

will be seen, we believe this is not a temporary phenomenon. Major changes have occurred over the past decade, with even greater changes likely over the rest of the century.

THE CURRENT ENERGY TRANSITION

The United States is today in the midst of a period of momentous transition—a historic and profound shift away from the use of liquid and gaseous fuels and toward a new and more diverse mix of energy sources. How did we reach this point in history?

As shown in Figure 1.1, energy transitions are not new. The United States has gone through two other energy transitions in the past 100 years. In the late 1800s there was a shift from wood to coal as the primary U.S. energy source, and in the early part of the twentieth century we shifted from dependence on coal to oil and gas. In the latest change, we are shifting away from oil and gas as our primary fuels. U.S. consumption of oil and gas peaked in the early 1970s at about 75 percent of total energy use, and has now started a decline.

Yet as we move away from oil and gas, something different is

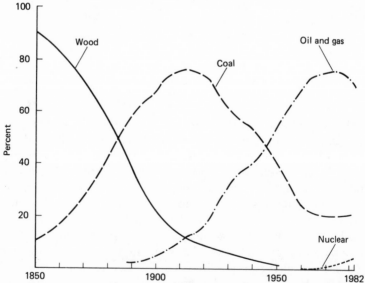

Figure 1.1   Major energy transitions, 1950–1982. (*Source*: Energy Research and Development Administration, *A National Plan for Energy Research, Development, and Demonstration*, U.S. Government Printing Office, Washington, D.C., 1975, p. S-2; with updated data from Energy Information Administration. *Monthly Energy Review* in August 1983.)

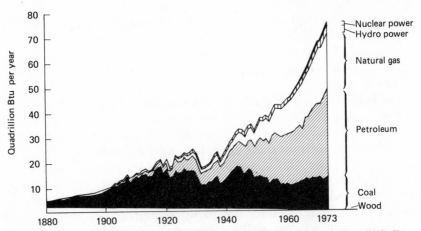

Figure 1.2   United States energy consumption, 1880–1973. (*Source*: U.S. Department of Commerce, *Historical Statistics of the United States, Colonial Times to 1970*. Washington, D.C., 1979, pp. 587, 588; and Energy Information Administration, *1980 Annual Report to Congress*, March 1981, p. 7.)

happening. In each of the previous two transitions, changes in energy use were caused by a surge of technological advances that took advantage of new energy sources. In each case, the new sources—first coal, then oil and gas—began to dominate our energy mix because they were cheaper and more productive than existing sources. Unlike previous movements, the decline in the availability of our current major sources— oil and gas—is not being accompanied by the emergence of a new energy source. This time, abundance is being created by energy productivity improvements—an extraordinary shift in the efficiency of appliances, automobiles, light bulbs, and the like—and it is these improvements that have made the current transition much smoother than anyone dared to expect.

In early energy transitions, the focus was entirely on one or another source of energy. For instance, in the nineteenth century, the industrial revolution was literally fired by coal. Coal was cheaper, more abundant and more convenient than the sources, primarily wood, that it replaced as shown in Figure 1.1. By the mid-1880s, coal was propelling railroad locomotives, heating homes, and powering industrial processes. In short order, coal supplied more than half the nation's annual energy needs.

By the early twentieth century, it was apparent that power from energy sources—offering the capacity to do work people once did—was becoming the motivating force behind the growth of a productive and vibrant new technological economy. Total energy use, as shown in Figure 1.2, grew exponentially through 1973 at a rate almost as fast as the economy. By 1900, wood's share of total fuel sources had slumped to 20 percent, and coal was nearing its peak as a fuel source, accounting for 70 percent of U.S. energy supplies. Oil and natural gas were just beginning to be used as energy sources, claiming less than 10 percent of the market.

As quickly as coal gained supremacy, however, it began to be replaced by liquid and gaseous fuels. The twentieth-century transition to oil and gas use was a technological wonder—abundant supplies of inexpensive, convenient and clean-burning oil and gas fueled a period of phenomenal growth in the U.S. economy. The development of our whole transportation system—cars, airplanes, trucks, and ships—was almost exclusively based on oil, and now oil or gas are the fuels required to satisfy much of our building and industry needs. After 1900, oil and natural gas steadily increased their combined share of the energy market at coal's expense to an astonishing 75 percent of U.S. fuel consumption in 1970.

Even though the dominant focus has been on energy sources, closely linked with each energy transition were major technological developments in energy-using devices resulting in improved engines, vehicles, light bulbs, motors, furnaces, boilers, and appliances. For example, when Thomas Edison invented the light bulb in 1879, he created a sudden demand for electricity, which soon became indispensable to the nation's modern lifestyle; light bulbs replaced oil lamps, and, later, electric motors replaced steam engines. Without advances such as the light bulb or the automobile, the seemingly easy fuel transitions could not have occurred, for without them, a market for each new fuel—first coal, then oil and gas—would not have been created. Building designs, industrial processes, and freight systems changed rapidly to accommodate the potential of each new fuel's economics and combustion characteristics.[1]

Taken together, the changing mix of fuels and end-use technologies caused widely different amounts of energy to be consumed for each dollar of economic activity. Overall, however, less and less energy was required per dollar of economic activity each year, as shown in Figure 1.3. In 1910, it took about 100,000 Btu to produce $1 of gross national product (measured in constant 1972 dollars). By 1973, the amount of energy needed had dropped 40 percent, to 60,000 Btu per dollar, a productivity increase averaging a little less than 1 percent per year. Yet

---

[1] As will become apparent, it is still the developments in these end-use areas and the choices that now exist in many of those fuel-using systems that enable us to assert that energy abundance—in the form of energy services—is being newly created.

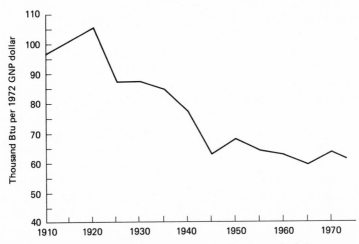

Figure 1.3   Trend of energy consumption in the United States per unit of GNP, 1910–1973. (*Source*: U.S. Department of Commerce, *Statistical Abstract of the United States (1979)*, p. 602.; U.S. Department of Commerce, *Historical statistics of the United States, Colonial Times to 1970*; U.S. Bureau of Economic Analysis, *National Income and Product Accounts of the United States*; and Energy Information Administration, *Monthly Energy Review*, April 1982, p. 2.)

over this historical period the downward trend was somewhat erratic, and this led to some misperceptions. For example, the data in Figure 1.3 show that the energy/GNP ratio was fairly constant in recent history (from 1945 to 1973), leading many to speak about the "lockstep" relationship between energy use and economic growth. From a longer-term perspective, however, the economy was becoming steadily more energy-efficient with these technological changes.

At the same time, the cost of America's energy, particularly electricity, was declining, as shown in Figure 1.4. Each new energy source could deliver energy services cheaper than the previous source, so the average cost of fuel dropped over most of our history. Cheaper energy and less energy needed per dollar of economic activity explains why past energy shifts occurred so painlessly and with so little public awareness. Each transition brought new, more efficient and convenient energy technologies, abundant and cheap new fuels, and increased prosperity. Everyone benefited from the changes. In 1970, there was little reason to suspect that this ever-improving energy situation would not continue indefinitely.

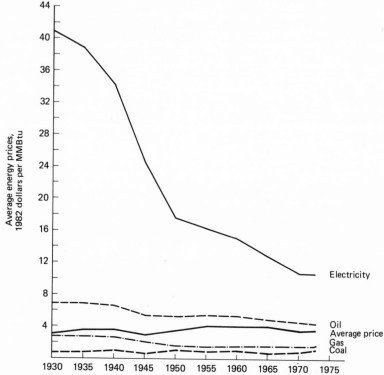

Figure 1.4   Historical energy prices. (*Sources*: Estimated from U.S. Department of Commerce, *Historical Statistics of the United States, Colonial Times to 1970*, Series M83-92 (Energy Consumption), E203-213 (Fuel Price Indexes); U.S. Dept. of Commerce, Bureau of Labor Statistics gasoline price index; and Energy Information Administration, *State Energy Price System*, vol. 1, 1982, *Monthly Energy Review*, February 1983, and *Annual Report to Congress* vol. 2, 1979. Electricity use and generation balances estimated from Energy Information Administration, *Form 4*; Edison Electric Institute, *Pocketbook of Electric Utility Industry Statistics*, 1983, p. 24, and *Statistical Yearbook of the Electric Utility Industry*.)

EARLY WARNINGS OF SHORTAGE

To be sure, even in the midst of this period of general contentment and boisterous economic growth, some people were concerned. From time to time, warnings of eventual energy scarcity were sounded. For instance, as long ago as 1908, before there was any perceptible sense of concern in the nation, President Theodore Roosevelt created a National Conservation Commission. At the Conference on the Conservation of Natural Resources (May 13, 1908) statements were delivered that, even though not specific to energy, have become articles of faith in many quarters today. Roosevelt himself stated: "It is time for us as a nation to exercise the same reasonable foresight in dealing with our great natural resources that would be shown by any prudent man in conserving and widely using the property which contains the assurance of well-being for himself and his children."[2]

Unfortunately, in an era of growth and prosperity, conservation was an unsettling subject for Americans, if not for the world at large. Later, Roosevelt extended his concern by requesting a meeting of the world's leaders to consider greater conservation of natural resources on a global scale. However, domestic politics and the election of William Howard Taft intervened, and efforts to implement Roosevelt's ideas failed.

Writing in *A Sand County Almanac* much later, in 1949, the legendary conservationist Aldo Leopold echoed the same sentiments: "There is as yet no ethic dealing with man's relation to land and to the animals and plants which grow upon it."[3] And, in 1952, the Paley Commission on Materials Policy, established by President Truman, recommended a national policy to develop alternate energy sources to avert a future fossil fuel shortage.[4] More than 2 decades later, a report to the Club of Rome, *The Limits to Growth*, suggested that civilization was at a critical juncture:

> Man can still choose his limits and stop when he pleases.... The alternative is to wait until the side-effects of technology suppress growth themselves or until problems arise that have no technical solutions. At any of those points the choice of limits will be gone. Growth will be stopped by pressures that are not of human choosing, and that . . . may be very much worse than those which society might choose for itself.[5]

Was there any evidence of an impending energy scarcity before the embargo? The answer is clearly a yes, but few people paid any attention. Most notably, M. K. ("King") Hubbert, a renowned petroleum geologist, theorized as early as 1950[6] that a continuation of the historical decline in oil and gas discovered per foot of exploratory drilling would eventually halt the growth in oil and gas production (the data are shown in Figure

[2] Albert B. Hart and Herbert R. Ferleger (eds.), *Theodore Roosevelt Cyclopedia*, Stanford Press, New York, 1941, p. 103.

[3] Aldo Leopold, *A Sand County Almanac*, Oxford University Press, 1949, p. 203.

[4] Materials Policy Commission, *Resources for Freedom*, U.S. Government Printing Office, Washington, D.C., 1952.

[5] D. H. Meadows, D. L. Meadows, J. Randers, W. W. Behrens III, *The Limits to Growth*, Universe Books, New York, 1979.

[6] M. K. Hubbert, "Energy from Fossil Fuels," *Annual Report of the Smithsonian Institution*, June 30, 1950, pp. 255–271.

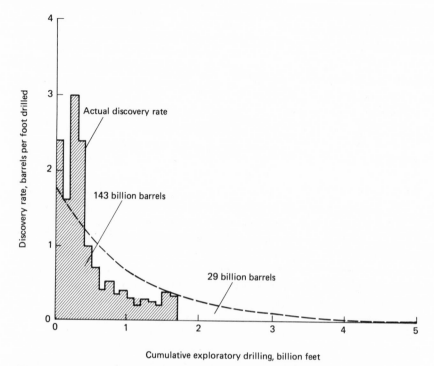

Figure 1.5 Oil discoveries per foot of exploratory drilling. (*Source*: M. K. Hubbert, *United States Energy Resources, A Review as of 1972*, prepared for the Senate Committee on Interior and Insular Affairs, pursuant to Senate Resolution 45, 93d Congress, Serial No. 93-40 (92-75), 1974, p. 125.)

1.5). According to Hubbert, oil and gas production must follow a "life cycle" similar to those shown in Figure 1.6—a period of low prices and exponential growth in production, a peaking of production as the effects of resource depletion cause drilling returns to drop and prices to rise, and a long period of decline in production.

But Hubbert's evidence was masked by the emergence of another seemingly cheap energy source—foreign oil. As U.S. oil and gas production weakened, the country became more and more dependent on foreign oil, which was imported in ever-increasing quantities to make up the difference between U.S. consumption and production (Figure 1.7). U.S. oil imports rose from 0.6 million barrels per day in 1950 to 6 million barrels per day in 1973, 35 percent of all oil consumed in the United States in that year. Although our ability to produce oil was limited, our ability to consume it seemed limitless. Lacking the domestic supplies to satisfy our needs, we shifted to foreign oil as a cheaper substitute for U.S. energy.

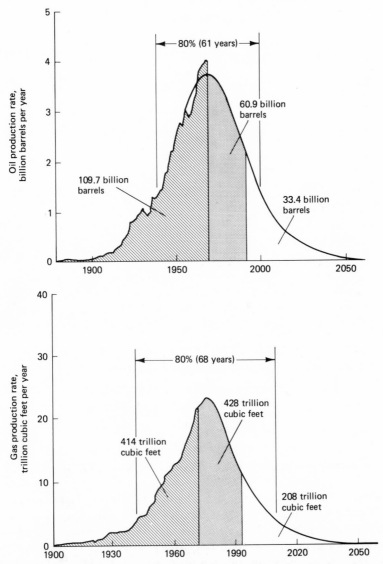

Figure 1.6 Hubbert oil and gas production life cycles. (*a*) Petroleum liquids in United States and adjacent continental shelves, excluding Alaska; (*b*) natural gas in United States and adjacent continental shelves, excluding Alaska. (*Source*: M. K. Hubbert, op. cit., pp. 152, 198.)

In retrospect, our collective national difficulty in dealing with energy is curious. It was always either feast or famine, scarcity or abundance, shortage or glut. A third alternative was seldom considered, namely that the real "energy crisis" was the inevitable transition from cheap to

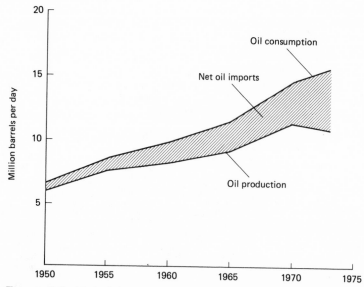

Figure 1.7 United States oil statistics. (*Source*: Energy Information Administration, *Monthly Energy Review*, August 1983.)

expensive energy. Even if the situation had been presented in that way, it is doubtful that, in its euphoria over a rising standard of living, the nation would have cared.

THE EMBARGO

With the violence of a summer thunderstorm, energy markets were suddenly disrupted by the Arab oil embargo of 1973–1974. Apathy was transformed into fear and panic. Suddenly the small phalanx of people who had been sounding the trumpet over the looming energy problem was joined by millions of angry citizens, all unnerved and enraged about gasoline shortages and skyrocketing utility and oil company bills. Because of the steep rise in all energy costs, caused by OPEC's ability to take advantage of the rising demand for imported oil by increasing its price, the decade of the 1970s saw the end of euphoria and the arrival of a set of grim new realities. The energy joyride was over.

As a result of this new concern, citizens across the United States demanded action. But it was soon evident that the general perception of the nation's energy predicament was flawed. Americans were told (or led to believe) that the energy problem involved only temporary supply interruptions and capricious Arab price increases, and, like the subsequent water shortage of 1976 in California, it would fade into memory if the citizens responded. Therefore, the solution was to be found in acts of patriotism to conserve energy by curtailing activities and in government controls to hold down the price of energy.

The principal advocate of that view was President Richard Nixon. Shortly after the embargo ended in 1974, he called for Americans to "pledge that by 1980 . . . we shall be able to meet America's energy needs from America's own energy resources."[7] The possibility that Arab oil policy might not be that different from policies of friendly suppliers, such as Mexico and Canada, was rarely, if ever, addressed. For political reasons, the spotlight was focused on OPEC. Not until the United Kingdom began charging more per barrel than Saudi Arabia did the realization spread that a much more profound, long-term energy transition was unfolding—one that could not be so easily blamed entirely on the Arabs or OPEC. The era of cheap oil and gas was ended.

In retrospect, the 1973–1974 embargo and consequent petro-shocks revealed that America's storehouse of cheap fuel resources was not as limitless as generations of Americans had been led to assume. When the market price for oil began to reflect the true cost of the next best energy supply alternatives, these prices were much higher than U.S. experts had predicted. A higher-cost energy future seemed unavoidable.

Meanwhile, Project Independence—the name chosen for the energy policies put in place in response to the 1973 oil embargo—resulted in a boomerang effect in the marketplace. Instead of declining, U.S. oil imports rose by 40 percent from 1973 to 1977. Even with the benefit of hindsight, it is difficult to recall a 4-year period during which the nation pursued such an unproductive policy course. What went wrong? Had the energy problem been analyzed incorrectly? Was scarcity the wrong theme? Had the appeal to patriotism, morality, and ethics been a major mistake? Is it conceivable that no one really understood the workings of the national energy system or the implications of the transition sweeping through the economy?

Part of the answer is that between the 1973 embargo and 1980, the United States underwent such an unprecedented change in circumstances that it was difficult for energy markets to respond quickly. In 1972 direct annual energy expenditures had declined to only 7 percent of the GNP, or about $960 per year per capita in 1982 dollars (see Figure 1.8). By 1980 the figure had more than doubled to over $2000 per person (in 1982 dollars), or about 15 percent of the GNP, largely because of an increase in oil costs. The declining energy-cost trend the nation had experienced for so many years was suddenly reversed.

---

[7] Federal Energy Administration, *Project Independence Report*, U.S. Government Printing Office, Washington, D.C., November 1974, p. 18.

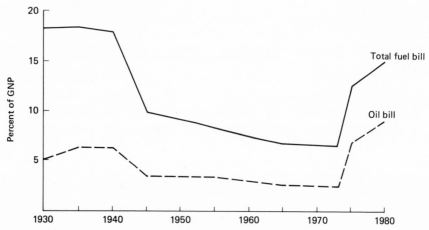

Figure 1.8   Total delivered energy bill as a percent of GNP. (*Source*: U.S. Department of Commerce, *Historical Statistics of the United States, Colonial Times to 1970*, and Energy Information Administration, *Monthly Energy Review*, August 1983.)

The nation's systems for supplying energy services had become badly imbalanced. Major investments in better furnaces, more insulation, and other energy efficiency improvements were needed to adjust to a new era of higher energy prices. Yet such an adjustment would take time— time for the energy-using stock of homes, cars, and industrial equipment to be replaced or refitted, and time for new energy supplies to be developed.

Yet in 1973 government policymakers were not willing to wait for the energy problem to solve itself. In fact, government officials were leery of exploiting economic self-interest by way of the market in choosing remedies to the energy problem. The result was a regulatory approach that emphasized immediate relief of the symptoms rather than an attempt to address the underlying longer-term issues driving the energy transition. The principal components of the government policy response was a decision to continue controlling the price of oil (just as natural gas prices were controlled in 1954 in a similar attempt to protect the consumer), accompanied by regulations on energy use and appeals and exhortations designed to conserve energy. While a momentary benefit was created in the sense that the U.S. protected itself from the damaging consequences of world oil prices, the Project Independence policies worked to impede rather than aid our transition from dependence on foreign oil. Because the economy was insulated from market pricing of oil and gas for so long, the public took longer to buy products and services that were available years ago to reduce their energy require-ments. It was not in the public's self-interest to buy new, more expensive energy-efficient products and services when prices were controlled at artificially low levels.

For an illustration of this delay, one need look no further than the sad saga of the U.S. auto industry. It took 5 years for domestic high-mileage auto sales to start up in earnest—a time when partially decontrolled gasoline prices more accurately reflected what the rest of the world was paying. For the same reason, homes and office buildings got a late start in being constructed in an energy-efficient way. It wasn't till the 1980s that major investments to reduce energy consumption were made, but the benefits are already visible, not just in the form of reduced energy bills but in greater comfort too.

ETHICS VERSUS SELF-INTEREST

Rather than trusting the underlying motivation of higher prices, there was a powerful tendency among our representatives to rely on ethical and moral appeals. One of the fundamental energy messages transmitted to Americans during the 1970s was that each citizen, for the sake of the nation, the planet, and posterity, had a moral responsibility to change his or her wasteful ways. Designed to prick the national conscience, these statements were the rhetorical descendants of the Roosevelt Commission credos, Aldo Leopold's writings, the Club of Rome conclusions, and others.

The answer lay in creating a new ethic. It was suggested again and again that society could not continue to operate on the arrogant conviction that natural processes can be fully understood or that the results of human intervention into nature can be predicted. It was pointed out that this conceit had produced dangers that could not have been predicted, probably cannot be corrected, and might eventually destroy the most "humanistic" of modern societies. And it was believed that unbridled self-interest would inevitably lead to a depletion of our energy resources and an assault on our planet.[8]

The ideals expressed were unassailable. Yet, despite the rhetoric and the political hurly-burly, price controls were driving America in precisely the opposite direction, encouraging continued energy consumption and forestalling investments in energy efficiency.

At the same time our political leaders—former presidents Nixon, Ford, and Carter—spoke to Americans of a different but related problem requiring a patriotic responsibility; the need to extract our-

---

[8] For a more current view from a group of the prominent environmentalists, see A. B. Lovins, L. H. Lovins, F. Krause, and W. Bach, *Least-Cost Energy—Solving the CO_2 Problem*, Brick House Publishing, Andover, 1981.

selves from the insecurity of undue dependence on foreign sources of oil. But such appeals elicited cynicism, inaction, and even ridicule. When President Carter called for individual sacrifices in the context of what he described as the "moral equivalent of war," his program soon acquired the nickname "MEOW." The energy situation called for a major reexamination of our society, but our energy policies were based on economic principles that worked against such a reexamination. Because of price controls, economic incentives and patriotic responsibility were firmly fixed as opposites. Low prices for oil and gas encouraged the use of more energy and more oil imports, while at the same time we were told our ethical responsibility was to use less. Given this conflict, people chose self-interest[9] over ethical responsibilities, and oil imports went up, not down.

This unfortunate outcome was foreseen by Sigmund Freud, who wrote in 1930 that "people have at all times set the greatest value on ethics, as though they expected that it in particular would produce especially important results."[10] But he said further that in making these ethical demands the culture "does not trouble itself enough about the mental constitution of human beings. It issues a command and does not ask whether it is possible for people to obey it. On the contrary, it assumes that man's ego is psychologically capable of anything that is required of it."[11] He concluded by predicting that if ethical demands go beyond certain limits, no matter how justified those demands are, a revolution will surely follow.[12] He stated that it is "quite certain that a real change in the relationship of human beings to possessions [e.g., energy] would be of more help in this direction than any ethical commands."[13] Applied to the energy situation, his words appear to mean that a "real change" in the price or the availability of energy may be more useful than a change in our ethical perceptions.

Reinhold Niebuhr, the classic liberal and sometime socialist philosopher, perhaps more accurately foretold that our mistake would not be in questioning whether self-interest was the same as our public and social interest, but rather in not using self-interest as a tool to achieve the socially desirable solution.[14]

---

[9] Meaning that they bought energy services they could afford at the lowest cost they could find. In this sense, their "self-interest" dictated that they could not afford more insulation or a higher-mileage car because their fuel bills were not lowered enough to make those actions worthwhile.

[10] Sigmund Freud, *Civilization and Its Discontents*, W. W. Norton, New York, 1961, p. 100.

[11] Ibid., p. 101.

[12] Ibid., p. 101.

[13] Ibid., p. 102.

[14] Reinhold Niebuhr, *The Children of Light and the Children of Darkness*, Scribner's, New York, 1960, pp. 9–15.

THE MOVE TOWARD MARKET SOLUTIONS

Toward the end of the 1970s, the public experiment with regulation was seen as a failure. U.S. oil imports remained intolerably high, despite President Nixon's Project Independence, President Ford's Whip Inflation Now, and President Carter's Moral Equivalent of War. There was no evidence that the multitude of regulations designed to replace price as the motivating force in adjusting energy markets had accomplished anything.[15] Out of desperation, public policy began a 180° shift—toward market rather than regulatory solutions.

In 1979, President Carter chose to decontrol oil prices over a 2-year period (President Reagan finished the deed in January 1981 by decontrolling all oil prices as of that date.) To the surprise of most, oil price decontrol worked much more effectively than expected, especially because the effects of the oil price increase arising from the overthrow of the Shah of Iran were so quickly felt. U.S. oil demand dropped 20 percent from 1978 to 1982. Oil imports were almost halved, from a peak of 8.6 million barrels per day in 1977 to 4.5 million barrels per day in 1982. This dramatic decline in demand forced world oil prices to drop more than 25 percent from their peak in the early 1980s, as the oil market sought a balance between supply and demand.

One might say that this was exactly what Adam Smith, the nineteenth-century guru of free-market advocates, had advised. Smith is often quoted as saying that when a man is guided by self-interest he is also "led by an invisible hand to promote an end which is not his intention."[16] That seems to be what happened in 1979–1980. People's self-interest reduced oil imports, but that was not their intention. But Smith also warned that sometimes people would have to sacrifice their individual interests to the public interest. He observed: "The wise and virtuous man is at all times willing that his own private interest should be sacrificed to the public interest of his own particular order of society— that the interests of this order of society be sacrificed to the greater interest of the state."[17]

As the nation is discovering with regard to energy, self-interest and the "public interest" are remarkably more compatible than we thought. But Smith's advice would have us go one step further—to reserve the right to change self-interest incentives if these interests alone do not achieve the greater interests of the state.

---

[15] In fact, a government study (U.S. Department of Energy, Energy Information Administration, *Energy Program/Energy Markets,* July 1980) concluded that the net impact of 6 years of public policy designed to combat the "energy crisis" placed the United States in about the same position as would have happened if no policies had been implemented.

[16] Adam Smith, *Wealth of Nations,* University of Chicago Press, Chicago, 1976, p. 27.

[17] Ibid.

At the time of the embargo, most people considered a reliance on market forces socially irresponsible. Yet today, in communities, businesses, and households, motivations of the market are being translated into social gains. Retrospectively, economic self-interest has proven to be the principal thrust behind the real progress being made in energy. Our articulation of this approach is a "Least-Cost Energy Strategy," where self-interested producers compete to deliver energy services to consumers at the lowest possible cost. We believe this de facto strategy is succeeding where other, more contrived strategies have failed in reducing the negative economic, environmental, and political impacts of the energy transition. But we also believe as Adam Smith did that we should carefully monitor the results of this approach in terms of the "greater interest of the state."

In the following chapters we will describe some of these innovative marketing approaches, technologies, and innovative individual and community actions which are producing exciting results in social and public terms. Then we will suggest a policy to guard against any future divergence of self-interest actions and national goals.

2

The Concept of Energy
Services

*The reason I wanted to sell light instead of current was that the public
didn't understand anything about electric terms or electricity.*

Thomas Edison

In 1977 two of us established the Energy Productivity Center of the
Mellon Institute, a division of Carnegie-Mellon University, to evaluate
the potential for increasing America's energy productivity. The principal analytic departure of the center from more conventional approaches
was to consider energy as a means of providing the services associated
with healthy economic growth. The concept of energy services is
different from traditional thinking about energy, and it is perhaps the
key concept of this book. It also provides the framework for our
assertion that energy abundance is being created once again.

The first public discussion of energy services was by Amory Lovins in
his classic 1976 article in *Foreign Affairs* magazine.[1] While he did not use
the words "energy services," he observed that for 220 million Americans,
energy is identified in terms of the services it provides: heat, light, and
mechanical motion. In a more prosaic sense, a comfortable room, a good
reading light, heat from a stove, or a seat in a moving vehicle are services
consumers require. In industry, managers have a similar preoccupation.

[1] A. Lovins, "Energy Strategy: The Road Not Taken," *Foreign Affairs*, vol. 55, October
1976, pp. 65–96.

17

The need is for heat to form steel, or shaft power to run machines, or steam for a distillation process. Energy is not an end in itself but only a means to these other ends.

Yet those of us who have been meshed in the politics of energy traditionally have looked at the issue not in terms of these services but in ways to manipulate the supply or restrict the consumption of fuels—how to manage British thermal units (Btu), kilowatt-hours, or barrels, instead of how to deliver heat, light, or motion. In short, we have been looking through the wrong end of the telescope—from the supplier's viewpoint, rather than the customer's. This book examines the behavior of energy markets from the customer's perspective, which starts with energy services.

ENERGY SERVICES

Figure 2.1 shows 1980 U.S. energy use according to three different accounting methods: *primary* fuel consumption, *delivered* or *end-use* fuel consumption, and consumption of *energy services*. In 1980 the United States consumed 78 quadrillion Btu (quads) per year of primary energy inputs, representing the original energy content of the coal, oil, gas, nuclear power, or renewable energy that was extracted to meet 1980 consumption. Of this amount, only 60 quads of fuel (oil, gas, coal, electricity, renewables) were delivered to residential, commercial, industrial, or transportation consumers. Some energy—18 quads, or 23 percent of the total primary energy inputs—was lost in converting primary energy to more usable forms (mostly in electricity generation) and in transmitting and distributing the energy to end-use customers.

Most people think of energy in terms of delivered British thermal units of fuel, for this is the way it is currently sold. The 60 quads of delivered fuels include the Btu equivalent of all the utilities' electricity and gas sales, sales of petroleum products, and coal sold directly to industrial (and some buildings') customers. Yet the energy in these fuels must be converted into energy services—more useful forms of energy— before they provide any value to the customer. This conversion occurs in energy-using devices in buildings, in industrial equipment, and in the rolling stock in the transportation sector. By our accounting, these energy services provided totaled 33 quads in 1980, only about 42 percent of the primary energy inputs in that year. Most of the missing energy was lost in converting fuels into energy services; for example, energy is lost in furnaces in providing heat to a building, in a boiler that provides steam to an industrial plant, or in a car or truck that moves people or freight.

Figure 2.1 also gives an accounting of the different categories of

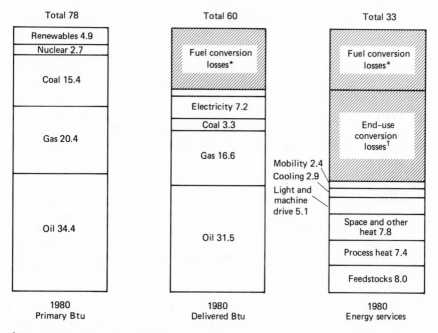

* Primarily electricity converson losses.
† Losses in converting fuel to energy services in end-use equipment.

Figure 2.1   United States fuel and energy services in 1980 (quadrillion Btu per year).

energy services consumed in 1980. Heat was by far the largest service category, comprising almost half of all energy services. About half the heat was used in buildings, primarily as space heat, and the rest in industry, primarily as steam for industrial processes. Petroleum-based feedstocks, which are used to produce chemicals, plastics, fertilizer, asphalt, and road oil, represented almost 25 percent of 1980's energy services. Cooling, light, and power for motors and appliances totaled another 25 percent. Because only about 10 percent of the energy content of transportation fuels are converted into mobility services (most of the energy is spent moving the carrier, not the passenger), mobility accounts for less than 10 percent of all energy services.

The nation's preoccupation with fuels, not energy services, has created misconceptions about energy demand. When speaking of energy demand, most reports focus on fuel use, not the demand for energy services. Yet it is energy services—not fuels—that people use. These services are a major input to our personal comfort and our economic activity. Even the inventor, Thomas Edison understood this difference, for he originally sold energy services, not fuels. Initially the service he provided was light. It was purchased as an amount of illumination on the basis of light-hours. The time the light was on was measured and billed at the equivalent, in today's dollars, of roughly 1 cent per lamp-hour.[2]

When the electric motor created more services from electricity, the New York Edison Company decided in 1898 to end the sale of light services and adopt the practice that is known today—the selling of electric current in terms of kilowatt-hours. This change incensed Edison, who explained later that "the reason I wanted to sell light instead of current was that the public didn't understand anything about electric terms or electricity." But more significantly, for our purposes, he went on to say:

> Another reason I did not want to sell current was that from my experiments, I knew that the incandescent lamp was only the beginning and that there were great possibilities of enormously increasing its economy . . . But for some reason the selling of current was introduced, thus destroying all chances of the company's gaining any benefits in [lamp] improvements; in fact, such improvements were a disadvantage, which in my mind is a poor business policy for the company and for the public.[3]

History has proven Edison right in many ways, but we are only beginning to understand the insights behind these statements. As a practical businessman, he knew it was foolish to give up the light bulb business. But from the perspective of consumers in the 1980s he was also suggesting that he didn't like the situation whereby it was a disadvantage for the electric utility to improve the efficiency of the light bulb—or implicitly, the efficiency of any other electric device.

THE COST OF ENERGY SERVICES

To deliver energy services requires more than fuels; the end-user must purchase, operate, and maintain a piece of energy-using equipment. Fuel costs are therefore only part of the cost of energy services—in many cases, a surprisingly small part. Table 2.1 shows that fuel costs were only about one-third of total energy service costs in 1972; the rest was for furnaces, motors, boilers, and other appliances, and fuel-using equipment. By 1980 fuel costs had increased to over half of total energy service costs because of the rapid increase in fuel prices over the decade.

A calculation of the dollars spent per capita in 1980 to provide

---

[2] Gerald S. Leighton, "The Electric Utility Industry: A Study of a Public-Private Relation," unpublished paper, George Washington University, School of Government and Business Administration, Washington, D.C., May 2, 1979, p. 6.

[3] Quoted in Charles E. Nell, "Entering the Seventh Decade of Electric Power," *Bulletin of the Edison Electric Institute*, September 1942.

TABLE 2.1 Components of Energy Service Costs

	1972		1980	
	1982 $/person	%	1982 $/person	%
Fuel	960	36	2095	51
Capital	1033	39	1223	30
Operation and maintenance	684	25	810	19
Total	2677	100	4128	100

SOURCE: 1972 estimates based on Oak Ridge National Laboratory, *Total Costs of Energy Services*, 1979, adjusted to conform with 1980 estimates. 1980 fuel, capital, and operating estimates from author's data.

different energy services also shows that the largest items are in some surprising categories (see Table 2.2). For instance, space heating and air-conditioning in residential and nonresidential buildings accounted for $629 of the average person's energy expenditure of about $4100, more than industrial manufacturing ($517) and feedstock ($103) expenditures combined. Yet it is industrial needs that are often thought of as *the* essential energy services to the economy. The numbers show further that the largest industrial expenditure—for electricity to drive machines and tools—averaged only $157 per person. These conclusions are not apparent when British thermal units are counted, suggesting that a consumer's view of energy—the cost of energy services—provides a very different and, we think, more useful perspective than a focus on fuel costs alone.

Other costs in Table 2.2 are more in keeping with our general perceptions: most of all energy services expenditures (about 60 percent) pay for mobility—getting people from place to place. And almost 60 percent of that is for car travel. Another 15 percent is spent to keep homes and offices comfortably warm or cool, and 10 percent is spent to run appliances and furnish hot water—a category we call convenience. Truck transport requires an expenditure close to the expenditure for keeping buildings warm. None of these findings is surprising, but what is surprising is that as a nation we do not account for our energy use in this way. Energy statistics are kept for fuels, not energy services. Statistics on energy costs therefore only account for a part, not the total, cost of energy services.

Table 2.2 also shows that the fuel cost portion of each energy service category varies significantly. For instance, fuel costs were only 36 percent of the cost of car travel, while fuel costs were 60 percent of the cost of hot water and thermal services in 1980. Obviously consumers could afford to pay quite a bit for an improved water heater to reduce the fuel cost of heating water—which is exactly what people are beginning to do.

TABLE 2.2 Estimated Cost of 1980 Energy Services in 1982 Dollars

Energy service	Total service cost	Annualized capital cost	Operation and main-tenance	Fuel cost	Fuel as % of total cost
Mobility:					
Car	1403	505	388	510	36
Light truck	394	117	122	155	39
Heavy truck	219	68	71	80	37
Air	106	23	24	59	56
Miscellaneous	268	98	102	68	25
Total	2390	811	707	872	36
Comfort:					
Space heating*	452	148	36	268	59
Air conditioning	177	51	12	114	64
Total	629	199	48	382	61
Convenience:					
Lighting, appliances	273	104	—	169	62
Hot water, other thermal†	144	57	—	87	60
Total	417	161	—	256	61
Manufacturing functions:					
Machine drive	157	8	9	140	89
Process steam	155	31	33	91	59
Indirect heat	80	5	5	70	88
Direct heat	67	4	4	59	88
Other process heat	27	3	3	21	78
Electrolysis	31	1	1	29	94
Total	517	52	55	410	79
Industrial feedstocks	103	—	—	103	100
Other‡	72	—	—	72	100
Total services	4128	1223	810	2095	51

* Including both industrial and building space heat.
† Cooking, dishwasher, washer/dryer.
‡ Primarily road oil, asphalt, and commercial gasoline use.
SOURCE: Estimated from Applied Energy Services, Inc., data.

A LEAST-COST ENERGY STRATEGY

Considering energy services rather than British thermal units of fuel was a breakthrough for us in how we approached the energy problem. The energy services concept forced us to consider energy in a broader context. Energy services can be provided by many competing means. Focusing on the economics of delivering energy services to consumers, we developed what we have previously called a Least-Cost Energy Strategy. Simply stated, this strategy has the goal of providing consumers with all the heat, light, and mechanical motion—or energy services—they demand at the least possible cost.

Within this broader context, there is a great deal of potential competition to provide energy services. The competing elements are the various fuel choices (oil, coal, natural gas, electricity, or renewables) versus the various methods of using or saving energy (efficient furnaces, insulation, fuel-efficient automobiles, and cogeneration). For example, oil faces stiff competition from energy-efficient engines. The interplay among the larger number of competing elements—all fighting to gain a share of each energy service market—is now providing energy services with less energy consumption than we have been accustomed to in the past. This enhanced competition requires an economic environment in which competition is unhindered by government interference—a path which for the most part we are on today. In fact, we now feel that the phrase "Least-Cost Energy Strategy" accurately describes the process by which Americans are mastering the energy problem.

To illustrate the Least-Cost concept, we prepared a hypothetical case for 1978 comparing the actual energy use patterns with what might have resulted if energy supply and end-use equipment were reconfigured to minimize consumer costs. More explicitly, the hypothetical 1978 case was our estimated response to the question: How would the nation have provided energy services in 1978 if its capital stock had been reconfigured to be optimal for actual 1978 energy prices? The results are summarized in Table 2.3.

Significantly, but not surprisingly, the difference in "mix" in the hypothetical case would have produced a different ultimate consumer cost. We estimated that the actual 1978 per capita cost of energy services

TABLE 2.3 Actual and Hypothetical Energy Use Patterns in 1978 (Share of Energy Service Market)

Energy resource or technology	% of total fuel use	
	Actual	Hypothetical
Oil	36	26
Natural gas	19	21
Coal (nonelectric)	4	3
Electricity (purchased)	30	17
Other	1	1
Improvements in energy efficiency*	10	32
Total	100	100
Cost per capita[†]	1146	984

* Efficiency-improvement technologies were compared to a 1973 base year.
[†] The cost of energy services, measured in 1978 dollars per capita.
SOURCE: These results were published in Roger W. Sant, *The Least-Cost Energy Strategy*, Energy Productivity Center, Arlington, Va., 1979, pp. 27–29.

(in 1978 dollars) was $1146. In the hypothetical case, that cost would have been $948. In other words, application of Least-Cost principles in the years prior to 1978 could have resulted in a 17 percent reduction in consumer outlays for 1978 energy services and proportional reductions in prior years.

Furthermore, the pattern of energy consumption in the hypothetical case would be considerably different from the pattern of consumption actually experienced in 1978. The overall shares of the energy service market captured by oil, industrial coal, and purchased electricity would be considerably smaller than they actually were: 28 percent smaller for oil, 30 percent smaller for coal, and 43 percent smaller for purchased electricity. A socially desirable side-effect of the hypothetical Least-Cost case would be a 50 percent reduction of imported oil.

A significant portion of this reduced consumption would result from improved efficiency. As a result of these energy savings, natural gas would increase its market share in the hypothetical case, and its use would be reallocated within various economic sectors. For example, gas consumption in the industrial sector would rise from 22 percent to 37 percent, permitting a substantial reduction in consumption of oil from 18 to 11 percent. A dramatic reduction of industrial purchased electricity from 26 to 11 percent would result largely from the substitution of less costly and more efficient electricity cogenerated on site.[4]

In the buildings sector, improved efficiency would have accounted for substantial reductions in oil, gas, and purchased electricity, with the gas becoming available for the industrial sector. Only in the transportation sector, where there is far less opportunity to substitute fuels, would no appreciable reallocation have occurred. Although the potential reallocation of shares of the energy service market among different fuels was interesting, the most important point to emerge from this exercise was the extent to which, in response to a Least-Cost strategy, technologies that improve energy efficiency might have captured shares of that market from fuels.

The potential gains in energy efficiency from a Least-Cost strategy would vary considerably in the industrial, buildings, and transportation sectors. For example, the gain in share of the market captured by improved efficiency in the hypothetical case would be considerably less in the industrial sector than in the buildings sector, owing largely to the fact that industry was able to react more quickly to energy price increases after 1973 and therefore had already taken greater advantage of opportunities for improved efficiency by 1978.

Taken together, technologies that improved energy efficiency would, in our hypothetical case, capture 22 percent more of the energy service

---

[4] "Cogeneration" is the simultaneous production of electricity and usable heat.

TABLE 2.4 Least-Cost Energy Efficiency Improvements for 1978

Efficiency improvement	% of added share of 1978 service market*
Structural improvement in residential buildings	5.3
Automobile weight reduction and power-train improvement	3.2
Improved space-conditioning equipment in buildings	2.0
Gas turbine cogeneration (industry)	1.6
Variable-speed motors (industry)	1.1
Appliance improvements	1.1
Structural improvements in commercial buildings	1.0
Automobile diesel engines	0.9
Light-truck weight reduction and power-train improvement	0.8
Cogeneration in buildings	0.7
Improved aircraft	0.4
Aircraft deregulation/load factor	0.4
Truck diesel engines	0.3
Heat recovery (industry)	0.3
Truck deregulation/load factor	0.2
All others	2.7
Total	22.0

* Additional share of 1978 energy services in the Least-Cost case, compared to actual 1978 market share.

SOURCE: Roger W. Sant, *The Least-Cost Energy Strategy*, Energy Productivity Center, Arlington, Va., 1979, p. 36.

market than they actually did capture in 1978 (the difference between 32 percent savings in the Least-Cost case and the 10 percent actual savings in 1978, as shown in Table 2.3). In other words, these technologies would enable us to reduce our total energy consumption by over one-fifth, bringing the United States closer to the levels of energy consumption (per unit of GNP) that have been attained in most other industrialized nations. Although dozens of categories accounted for the 22 percent increase in new market share for energy efficiency in the hypothetical case, building improvements dominate the list. The largest single category—representing about 25 percent of the potential savings—was structural improvements in residential buildings. Building energy improvements total about half of the 22 percent additional savings shown in Table 2.4.

THE LIMITS OF THE LEAST-COST EXERCISE

The purpose of this analysis was not to suggest what exactly could have occurred in 1978 under the Least-Cost strategy. Because rising energy prices had only been a reality for about 5 years at that time, it would be unreasonable to expect that the United States could have achieved the optimum Least-Cost mix of the hypothetical case in so short a period. In addition, our analysis was based on actual 1978 prices, which might have been considerably different had competitive conditions prevailed during the preceding decade. The dynamics of a fully competitive market were and are difficult to predict. For instance, if the demand for oil had been as low as in the hypothetical case, the world, or OPEC, oil price would not have been as high as it actually was. Similarly, the price of electricity might have been different—either higher or lower—if the demand had been reduced to the level of the hypothetical case. And the price of natural gas would have been much higher if it were not controlled.

In addition, the Least-Cost results assumed the operation of a perfect economic market, and such a market is not likely ever to exist. Non-economic behavioral and institutional factors work to dilute our hypothetical results in the real world. On the other hand, in 1979 we had only begun our research effort to identify technologies that reduce the cost of energy services, and there were many opportunities which were not yet included in our data base or which had not yet been developed at the time of our analysis. The analysis was therefore purely hypothetical, intended simply to illustrate the Least-Cost concept.

The chief outcome of this exercise was to demonstrate how, through price controls and other means of intervention, the nation had veered away from the Least-Cost path—and to demonstrate that other, more productive directions were among our current options. It indicated theoretically that consumers paid roughly 17 percent more for energy services in 1978 than they needed to pay.

Our current stock of buildings, vehicles, and industrial equipment is an accumulation of years of investments. This stock takes years to turn over, buildings might typically last for 50 to 100 years, industrial equipment for 15 to 30 years, and vehicles for 6 to 10 years. Our current capital stock, in many ways outmoded and poorly matched to our current and future energy situation, represents a large business potential—a source of energy abundance to be tapped over the next few decades. Although achieving comparable results to those suggested by our hypothetical case might indeed take decades, the analysis did indicate the direction in which the country might move to begin realizing some of the benefits of a Least-Cost strategy. Many of those are now being utilized. In Chapter 7 we show some current results from using this same methodology. These forecasts, if anything, are more conclusive than the 1979 exercise about the effect of greater efficiency on U.S. energy demand. Before showing these results, we want to provide some illustrations of the Least-Cost, or energy services, phenomenon.

EXAMPLES OF THE SHIFT TO
ENERGY SERVICES

Perhaps the most notable example of a traditional energy organization moving toward a Least-Cost approach is an association of public utilities (e.g., the utilities that are owned by consumer, municipalities, or other public institutions), the American Public Power Association (APPA). The APPA has endorsed a concept they call "energy services planning." As Larry Hobart, the APPA deputy executive director, stated to a congressional subcommittee:

> Energy services planning is based on the belief that as consumer-owned entities, local public power systems have an obligation to supply their customers' needs for heating, cooling, lighting, and motors at the lowest possible cost consistent with sound business principles through use of the most cost-effective means, whether it is energy conservation, renewable resources, or central-station electricity.[5]

Within APPA, there are several examples of how member utilities are implementing the energy services or Least-Cost concept. The Easton, Maryland, municipal utility is helping its consumers meet their heating and cooling needs by providing water-to-air heat pumps which use 50° to 60°F city water as a heat source and heat sink. The Tennessee Valley Authority (TVA) has the most extensive residential retrofit program in the country. So far, 50,000 homes have been improved. Phillippi, West Virginia, is in the initial phase of a citywide community program to spread the residential demand for electricity more evenly throughout the day. Albany, Georgia, has almost 5,000 central air conditioners and water heaters under time-of-day control, again to spread the electric demand more evenly during the day. They have concluded that the entire system paid for itself in the first 60 minutes of use.

Massachusetts Municipal Wholesale Electric Company is recovering heat rejected from one of their power plants to provide space conditioning for its headquarters office building. Among the many cities that have initiated Least-Cost services projects are: Turlock, California; North Little Rock, Arkansas; Palo Alto, California; Burlington, Vermont; Eugene, Oregon; Lamar, Colorado; and Ames, Iowa.[6] Bonneville Power Authority is sponsoring 10 field tests of energy service company concepts.

---

[5] Larry Hobart, *Testimony to the Subcommittee on Energy Conservation and Power*, Washington, D.C., Apr. 23, 1982, pp. 1, 2.
[6] Ibid., pp. 36–38.

Some private or investor-owned utilities also have some impressive programs underway that exemplify the energy services approach. San Diego Gas and Electric has a subsidiary (recently spun off) that makes electricity and steam through cogeneration and sells the steam at a discount to customers, primarily the U.S. Navy. Pacific Gas and Electric provides zero-interest loans for their customers to improve the efficiency of their homes and hence lower the cost of heating and cooling. Northeast Utilities has an extensive program of residential conservation and load management. Southern California Edison, Pacific Power and Light, Southern California Gas, General Public Utilities, Public Service Company of New Mexico, Arkansas Power and Light, Florida Power and Light, Texas Utilities, and many others have programs of various kinds that are testing the energy services approach. In fact, a report by the Investor Responsibility Research Center in Washington suggested that a "virtual stampede" now exists to make conservation and load management—key elements of an energy service approach—part of their overall operations. About 86 of 120 utilities surveyed have adopted formal conservation programs.[7]

If utilities increasingly compete for market share in energy service markets—and, as we have shown, there are signs that some of them are taking such competition seriously—they will begin to move away from their historical position of a regulated utility to a more competitive market. The resulting competition will provide more choice, better quality, and lower costs to consumers. The possibilities for competition arise from the likelihood that increased investments in the use of energy service systems will likely be larger than for new fuel supplies. That is not because of anyone's moral or ethical ideas on what the world ought to be but rather because that's the direction in which economic self-interest is pushing us. Successful new retailers will be those organizations that can provide balanced systems for heating, cooling, and lighting, not just fuel. The systems include furnaces, insulation, automatic controls, light bulbs, and fixtures at the lowest total energy service cost.

In some instances, paying off the cost of all system improvements is becoming a part of a customer's lower monthly cost. Instead of requiring customers to balance the fuel, structural, and equipment costs themselves, an energy services company provides only those improvements that make economic sense and charges its users one low monthly price for the complete service—which includes the cost of fuel, a monthly charge for equipment, and a management fee.

A critical marketing shift is occurring—from a strategy of selling fuel

---

[7] Douglas Cogan and Susan Williams, *Generating Energy Alternatives: Conservation, Load Management, Renewable Energy and America's Electric Utilities*, Investor Reasonability Research Center, Inc., 1983.

or equipment to a view that treats these items as a means to provide complete energy services. Recalling the concept Ted Levitt wrote about in 1960, organizations that understand and meet the new energy services requirements are capturing the market and are more likely to capture the bulk of consumer expenditures in coming years.[8] Conversely, those that do not—which may include some of today's utility and oil companies—will become the "railroads" of tomorrow, and the new energy service organizations will become the "air express" companies of today.

Regardless of who gains in the struggle for market share, an investment of approximately $1 to $2 trillion for products and technologies that lower the cost of energy services is now justified.[9] Many nonenergy companies have begun to examine how their organizations' unique qualifications can capture a sizable share of those purchases. Wall Street analysts are talking increasingly about these markets and the likely beneficiaries. But the real beneficiaries of this competition are consumers—giving each energy user a broader range of new alternatives, which can put a cap on previously spiraling fuel costs.

It is clear that improvements in energy productivity that bring about these desirable results will also lead to overall productivity improvements for the economy. A whole range of companies have emerged with new developments and expertise in balancing energy systems. However, the most progressive energy companies in the next decade will probably be companies like Honeywell or Johnson Controls, not oil companies or utilities. The most exciting technology is likely to be something mundane like slow-speed coal slurry-fueled diesel engines that can cogenerate heat and electricity rather than synthetic fuels or breeder reactors.

Therefore, for the first time in 15 years there is reason to be optimistic about America's energy future. The challenge that needed to be met is being met. Even though it was late in starting, the job that needed to be done is being done through old-fashioned American ingenuity and competition. Some would say this is a romantic vision. On the contrary, it is a reality. The capacity that has been within our nation's grasp is being exploited by manufacturers, contractors, state utility commissions, bankers, building code jurisdictions, and, of course, all consumers. Yet the opportunities continue to be extraordinary. If anything is missing, it is the knowledge of how much, in fact, is going on.

[8] Theodore Levitt, "Marketing Myopia," *Harvard Business Review*, vol. 38, July–August 1960.

[9] Chapter 7 describes the nature of these potential investments in detail.

3

Mobility Options: A Creative Future

The nation need not fear that it will run on empty: in fact it could run on full and not use a drop of imported oil.

The most familiar use of energy service is to provide mobility—to move people and goods from one place to another. Mobility is central to our modern way of life; thus, it is not surprising that no other energy service has received more attention from politicians and economists. A few hundred years ago the majority of people lived and worked within walking distance of where they were born. They no longer do. This profound new arrangement places a premium on low-cost transportation modes.

Perhaps because the nation was caught unaware of its dependence on mobility in 1973, the fear of running on empty caused by the embargo created an atmosphere of insecurity and fear. For many citizens gasoline *was* the energy crisis, which is why transportation, especially via automobile, has received so much attention since oil prices started to climb and supplies were interrupted in the 1970s. Throughout this period, experts have been telling people that "the most irreplaceable energy is the liquid fuel required for autos and other transportation."[1]

Though it is tempting to accept this point of view (since most people

[1] For a discussion of this and what we consider to be other energy myths, see Chapter 8.

cannot imagine alternatives to their automobile or to gasoline or diesel fuels), our analysis has led us to a different conclusion. As in the other energy-consuming sectors of the economy, mobility is a service and can be expressed in passenger miles traveled, or ton-miles in the case of commodity shipments. As a part of the Mellon Institute research, Richard Shackson and H. James Leach made a detailed assessment of U.S. mobility options.[2] By concentrating on mobility (not fuels), and by expanding the focus of how that mobility is produced, they found an abundance of mobility options, not a shortage of oil. The analysis demonstrates that the United States need not be dependent on ever-increasing quantities of gasoline, as was the case in the 1960s and 1970s. In fact, it appears that mobility can and probably will be produced in bountiful quantities in the future, while relying less and less on the unpredictable supply of petroleum.

DEFINING MOBILITY SERVICES

In order to frame the case for this more abundant future, it is necessary to focus on the many different energy services which traditionally were lumped together under the category of transportation. Unquestionably, these services are difficult to categorize. In the people-moving category, for example, many kinds of services are required; trips to work, school, or shopping; trips for entertainment or recreation; trips for personal or family business; and business trips. Each kind of trip can be further categorized by length and the number of people and possessions traveling together. Then there is the question of what mode is selected for each trip, and within modes, what vehicle type is best suited or desired for a given trip. This same type of analysis is necessary in considering moving freight rather than people. Taking this approach, a very diverse and complex set of services materializes under the overall category of mobility.

The cube matrix in Figure 3.1 illustrates the varied nature of personal mobility services. For example, in a short trip for entertainment or recreation, the choices are car, bus, train, motorcycle, and bicycle. However, if the weather is bad or safety is a concern, riding a bike might not be an option. Similarly taking the bus on such a trip might be difficult if, for instance, one is going water skiing and must carry skis or fuel for the boat.

One way to understand the popularity of the family car is to examine each compartment in the cube and observe that the automobile is the leading option for providing most of the services required. It is the

---

[2] Richard H. Shackson and H. James Leach, *Maintaining Automotive Mobility*, Energy Productivity Center, Mellon Institute, Arlington, Va., Mar. 1, 1982.

choice of service for almost all medium-length trips and for most of the short trips as well. It is also the leading competitor in many of the long-trip entertainment and recreation compartments. In fact, 84 percent of all trips are made by car or light truck (Figure 3.2). The ability to provide so many different services at competitive costs is precisely why the auto is so central to our modern industrial society.

One surprising transportation statistic shown in Table 3.1 is the irrelevancy of mass transit in the mobility picture. Even if by some stroke of magic or government program mass transit ridership could be doubled or tripled in the next few years, its impact on total mobility energy services would be very slight, since it now accounts for less than 3 percent of all person-trips. Even so, mass transit is important to the quality of life in cities and plays an important role in urban transportation.

Even in the long trip or intercity travel categories as shown in Figure 3.2, the automobile is dominant. The share of services provided by airlines has increased dramatically, however, rising from 4.4 percent in 1960 to 14.2 percent in 1980. The volume of intercity passenger miles was over 1.5 trillion miles, out of a total of over 3 trillion passenger miles.

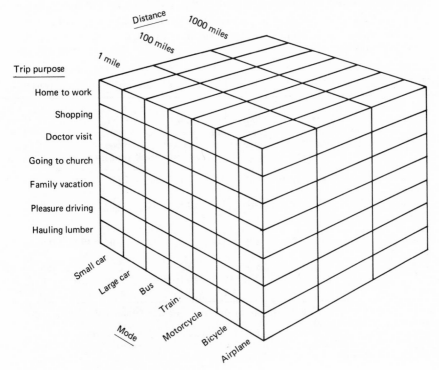

Figure 3.1 Personal mobility cube.

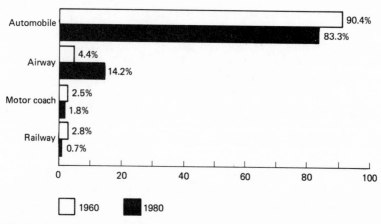

Figure 3.2 1960 ▢ 1980 ■

Figure 3.2 Percent of intercity passenger miles by mode of travel. (*Source: Motor Vehicle Facts and Figures—1981*, Motor Vehicle Manufacturers Association, Public Affairs Division, Detroit, Michigan, 1981.)

Transportation of freight is shared among trucks, trains, ships, and pipelines (see Figure 3.3). The largest growth in recent years has been in truck shipments. It has tripled since 1959 and is expected to continue upward. Even so, among all highway vehicles, including trucks, the automobile still dominates as the largest energy-consuming transport mode.

TABLE 3.1 Percent of Person-Trips by Trip Purpose and Mode of Transportation

Trip purpose	Auto and station wagon	Vans and pick-ups	Other*	Subtotal personal use	Bus and street car	Subway, elevated rail	Other†	Subtotal mass transit	Other‡	Total	Average trip length (miles)
Earning a living	74.0	13.2	1.9	89.0	2.8	0.7	0.9	4.3	6.6	100	9.6
Personal business	77.6	9.6	1.1	88.4	1.0	0.1	0.3	1.5	10.2	100	5.9
Civic, educational, and religious	49.8	4.0	0.3	54.2	4.6	0.3	0.1	5.0	40.8	100	6.1
Social and recreational	67.3	7.1	1.0	75.3	2.2	0.2	0.2	2.6	22.1	100	10.2
Other	78.7	7.9	0.9	87.4	2.4	0.7	1.4	4.4	8.1	100	9.8
All purposes	73.1	9.3	1.2	83.7	2.0	0.3	0.5	2.8	13.5	100	8.3

* Includes other trucks, motorcycle, self-contained recreational vehicle, and taxi (personal use).
† Includes train, airplane, and taxi.
‡ Includes bicycle, walk, school bus, moped, and other modes not elsewhere classified.
SOURCE: U.S. Federal Highway Administration, *1977 National Personal Transportation Study*, published in MVMA *Motor Vehicle Facts and Figures—1981*.

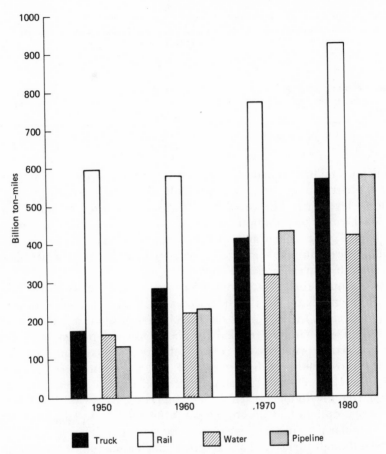

Figure 3.3 Intercity freight ton-miles by mode. (*Source*: *Motor Vehicle Facts and Figures—1981*, Motor Vehicle Manufacturers Association, Public Affairs Division, Detroit, Michigan, 1981, p. 49.)

THE COST OF PERSONAL MOBILITY

To understand the options for personal mobility, it is useful to focus first on the cost of driving an automobile, which is best expressed on a cost-per-mile basis. Table 3.2 analyzes all the elements of cost that go into a mile of travel by car. Surprisingly, fuel costs are only about 25 percent of the total cost of operating a car. Most of the dollars spent on personal travel are spent on the cost of the automobile (34 percent), maintenance and repairs (23 percent), and parking and insurance (17 percent).

Although the cost per mile of driving has risen over time, the personal expenditures for transportation by automobile as a percent of

TABLE 3.2 Automobile Energy Service Costs in 1980

Cost category	Cents per mile	% of total cost
Fuel	6.5	26
Cost of auto	8.4	34
Repairs and maintenance	5.6	23
Parking and insurance	4.3	17
Total	24.8	100

SOURCE: Applied Energy Services, *The Least-Cost Update*, op. cit., p. 127.

family income have remained relatively constant, at around 13 percent. On a per-trip basis autos are relatively cheap to operate, but on a national level their extensive use adds to a substantial sum. In 1980, personal auto transportation accounted for 1.5 trillion miles of travel in the United States. That amounts to total energy service costs of $380 billion, representing one-eighth of the GNP in 1980. About $100 billion was spent by travelers on fuel alone.

Even though fuel costs are only 25 percent of the total cost of driving, the rise in gasoline prices has received a great deal of attention during the past decade. In 1973 crude oil cost $3 per 42-gallon barrel, or 7 cents per gallon, yet gasoline retailed for about 50 cents per gallon in that year, reflecting primarily the cost of refining and marketing. By 1975 crude oil costs had risen five times, and the gasoline price had doubled. After the 1979 oil crisis, gasoline prices had risen to about $1.25 per gallon, roughly doubling each driver's fuel costs. By early 1983, however, gasoline prices had declined again to about $1.10 per gallon, with some regular gasoline selling as low as 90 cents per gallon.

Even though fuel costs have received much attention, we should not ignore the other $300 billion spent annually to provide automobile transportation services. Our annual expenses for repairs and maintenance almost equals the amount we pay for fuel, and parking and insurance adds about $70 billion per year. That still leaves $100 billion in annual capital expenses on vehicles. The secret to reduced mobility costs lies in the trade-offs between capital equipment costs and fuel and operating expenses.

The Shackson-Leach analysis identified over a dozen automotive technologies, some of which are described on the following pages, that can reduce the cost of driving. In each case fuel savings would more than pay for the added technology. Over the next 20 years these cost-effective mileage improvements should keep driving costs at about the same level as 1979. In the process, gasoline consumption (shown in Figure 3.4) might be cut in half, even with more drivers, more cars, and more travel. The potential savings from fuel economy are surprisingly large.

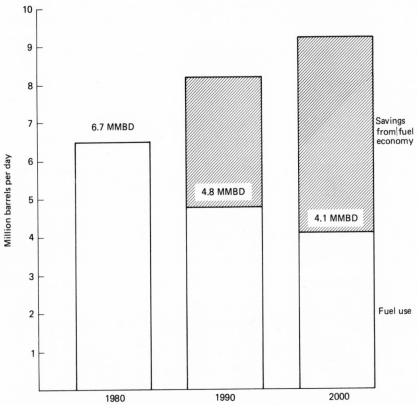

Figure 3.4   Fuel use for personal mobility (cars and light trucks). (*Source*: Applied Energy Services Least-Cost projections.)

MOBILITY OPTIONS WILL BE FOUND IN STRANGE PLACES

The abundance of fuel to operate cars will not come from oil fields far off shore or from synthetic conversion of coal or oil shale. It will more likely be the result of dramatic changes in new auto designs that reduce gasoline consumption. Here are some of the more visible changes that will occur in the auto industry during the next 20 years that could reduce the cost of driving.

Making Cars Smaller

Since 1973 downsizing cars has been a major part of the auto manufacturers' strategy to reduce fuel consumption. Typically, downsizing involves shortening the car and reducing its wheelbase. In many cases, this can be done without sacrificing the interior space of the car. Between 1980 and 1985 auto efficiency is likely to be improved from downsizing by 12 percent, an equivalent of finding an oil field that yields about 30 percent of the oil coming down the Alaska pipeline.

Weight Reductions

Parts of cars can be and are being redesigned so that they perform the same functions with less material. This reduces weight. By the year 2000 these design changes will probably reduce car weight by 100 pounds or more, thereby improving fuel efficiency. An even more important way to reduce weight is to make auto parts from different, lighter materials. This means the increased use of reinforced plastics and aluminum. These actions could reduce the average new car consumption of gasoline by 13 percent by the year 2000.

Aerodynamic Improvements

More improvements in reducing wind resistance are on the way. New vehicle designs and add-on equipment like underbody pans, front and rear spoilers, and streamlined mirrors will reduce the drag on cars, not only from head wind but from crosswind. Together these changes are expected to reduce gasoline consumption 6 percent.

Front-Wheel Drive

Front-wheel drive does not improve auto efficiency by itself, but it does allow for better use of the interior space of the car, thus reducing overall size, weight, and fuel consumption.

Reduced Rolling Friction

Enabling a car to roll more easily is being accomplished by improving tire design, increasing tire pressure, and reducing brake drag. These changes will improve fuel economy by 2 percent, on average.

Lubrication of Auto Parts

Improvements can be made in the oil and grease used in wheel bearings, transmissions, differentials, and the engines themselves. All of these changes will reduce gasoline consumption.

Improved Transmissions

One of the most significant areas in which technology can substitute for fuel is in improved transmissions. These new transmissions will have typical automobile labels such as "four-speed automatic with torque converter lock-up." In the 1990s we may even see electronically controlled transmissions that match engine speed and road speed more effectively, or even continuously variable transmissions that perfectly match the engine to road speed. Over the next 20 years, these changes should improve fuel efficiency 20 percent.

Improved Engines

Improved engines, including diesels, will continue to be a major focal point for auto efficiency. These changes include the turbocharging and supercharging of existing engines; improved operating characteristics of engines, particularly through electronic engine control; and fundamental improvements in combustion technology, heat transfer, and internal power losses. Diesels now offer a 25 percent fuel economy advantage.

In the 1990s other power-train options could be available to replace the conventional gasoline engine: electric cars, hybrid cars, gas turbines, Stirling engines, as well as clean, advanced diesels. In addition, new types of vehicles efficiently suited for their purpose—for example, city cars—may become popular. But all of these technical opportunities must withstand the same scrutiny before being deemed acceptable: they must provide mobility services at reduced costs to the consumer.

Together these technical changes will accelerate the dramatic change in auto fuel efficiency which commenced in the early 1970s. The average efficiency of a new car could easily reach 40 miles per gallon by 2000; it could go much higher depending on the future marketability of technologies now under development and the eventual cost of gasoline. Let there be no mistake, however; the cost of these improvements will be substantial. An estimated $6 to $9 billion in capital investment by the industry will be required annually for these fuel efficiency improvements, leading to a total expenditure of $88 billion by the year 2000. For consumers, that means cars might cost 20 to 40 percent more in the next two decades. Nevertheless, because these cost increases "purchase" reduced fuel consumption, the cost per mile, even with higher gasoline prices and more expensive cars, can be reduced 2 to 8 percent in the year 2000.

In this fashion, energy abundance is being created. A record for mileage improvements has already been established in the auto industry. In 1979, all cars and light trucks consumed 6.5 million barrels per day of gasoline. In the year 2000, with an estimated 20 percent increase in the number of cars and trucks, gasoline consumption should continue to drop to 3.5 to 4 million barrels per day as further transportation cost improvements are made. Therefore, it is really a myth to believe that the nation needs ever-increasing amounts of gasoline to preserve its mobility options.

IMPROVED TECHNOLOGY IN FREIGHT MOBILITY

The trucking industry has been growing rapidly since the early 1950s. Highway freight traffic expanded dramatically along with the interstate highway system in the United States. Truck transportation is convenient, fast, and reliable and offers virtually door-to-door service. The industry is expected to continue to grow and thus needs an increasing share of transport fuel supplies. Currently, trucks use about one-quarter of all transport fuels. However, the future for energy savings in trucking is bright. Most of the large semitrailer trucks are already diesel-powered; therefore they are already 20 percent more efficient than if they were gasoline-powered. Less than half of all medium-sized trucks are diesel-powered, so there is room available immediately for efficiency improvements in this category as new diesel trucks enter the market.

There are other developments in store for trucks similar to the technological improvements in personal automobiles. In the early 1970s people began noticing odd-shaped pieces of metal welded on the top of truck cabs. This was nothing more than a technique to lower wind resistance, and it was cheap. Almost every large truck on the highway today is so designed. In fact, further aerodynamic improvements in trucks could save a considerable amount of fuel—perhaps 10 to 15 percent.

There are numerous other ways of reducing fuel consumption in trucks; the use of lightweight materials, electronic engine controls, low-resistance tires, high-torque engines, and improved transmissions. The technology improvements may not yield quite as many savings as in automobiles. But there is one very effective option available to the trucking industry which will improve economy at low cost, and that is driver training. Drivers trained to optimize fuel economy have saved 20 to 30 percent in fuel costs, with no capital costs and only modest increases in travel time.

In addition, there has been considerable debate over the size and weight limitations that are imposed on the trucking industry. The industry argues that larger, longer, and heavier trucks will improve productivity, save fuel, and reduce shipment costs. On the other hand, trucks cause a great deal of wear and tear on the highway system, and it is believed that the trucking industry does not pay its fair share of construction and maintenance costs. The recently passed Surface Transportation Assistance Act of 1982, the so-called Nickel Tax Bill, addresses these very issues by raising the user fee for large trucks to help maintain the nation's roads, while at the same time allowing larger trucks on the highways to increase productivity.

In all, commercial trucks presently use about 2.4 quads of energy each year, equal to 1.2 million barrels per day of gasoline and diesel fuel. The services provided by the trucking industry are expected to expand nearly twofold by 2000, but the fuel consumption for trucks might rise only about 25 percent. Truck fuel efficiency could be 40 percent higher by 2000.

AIR TRAVEL COSTS

The prospect for more efficient air travel is also bright. Since 1973 fuel costs have risen from 25 percent of operating costs to nearly 60 percent for most airlines. This price shock has created major opportunities for large investments in technologies that reduce the cost of air travel and fuel consumption per passenger-mile traveled. For example, the much-acclaimed Boeing 767 is now on the market, along with its smaller partner, the 757. The 767 is the first completely new plane produced by the U.S. aircraft industry in a decade. Both new planes feature the latest in technological advances—quieter engines that use 35 percent less fuel, sophisticated computerized flight control and instrument systems, light-weight structural material, and a larger wingspan design. Another new cost-saving feature is the redesigned cockpit that needs only two crew members. So great is the potential for aircraft efficiency improvements that some experts are predicting that the next generation of planes will use half the fuel of the present fleet. If this future materializes, petroleum could once again become a relatively unimportant factor in the cost of flying. At present, however, slower-than-expected fuel cost increases and the glut of used aircraft are holding back the rapid adoption of new, streamlined planes in the marketplace.

Apart from new aircraft, there are other options for making air travel more cost-competitive. Between 1973 and 1979, passenger-miles per gallon of fuel increased by 50 percent.[3] This was accomplished by flying at reduced speeds, flying at more efficient altitudes (e.g., where wind

---

[3] Charles L. Blake, *The Impact of Petroleum, Synthetic and Cryogenic Fuels on Civil Aviation,* U.S. Department of Transportation, Federal Aviation Administration, June 1982.

resistence is less) and grounding inefficient airplanes. More recently, passengers have been spending less time circling around airports waiting to land. The new realities of expensive fuel prompted the practice that jetliners are not permitted to take off until a landing time is assured at their destination.

There is also evidence that air travel costs can be reduced substantially through greater use of automated air traffic control and flight management equipment. Over time the fuel cost savings from these improvements could be as high as 20 percent. Another profitable area that could produce an estimated cost reduction of nearly $300 million per year is through improving the existing aviation weather measurement and forecasting system. Currently, airline flights are often planned on old weather data. Thus, it is not surprising that planes encounter delays, diversions, and unfavorable wind and temperature conditions. As a practical matter, without improved weather measurement, cost savings achieved through the use of sophisticated data management systems, optimizing flight planning procedures, and maintaining the airplane at peak efficiency can disappear in a few minutes of unscheduled delay or in an unexpected headwind.

The oil crisis catapulted the airline and aircraft industries into a reassessment of their future. Now, a decade later, impressive strides have been made in combining cost-saving measures with financial prudence. Indeed, the options for reducing air transportation costs from their current high level are extensive. Our analysis suggests that the real costs for air travel could drop almost 20 percent by the year 2000, with a 50 percent increase in fuel efficiency.

SUBSTITUTES FOR PETROLEUM

Alternatives to petroleum-based fuels could play a role in powering the transportation system in the future. These substitutes for oil also increase the options in providing low-cost mobility services and are available if oil prices should rise to high levels. The leading candidates for future transportation fuels are methane (natural gas), propane, ethanol, methanol, coal liquids, oil shale, and electricity. All of the substitute fossil fuels mentioned are carriers of hydrogen, and ultimately pure hydrogen could even be used to provide a clean and efficient fuel.

The principal alternative transportation fuels are natural gas, biomass (a form of solar energy), oil shale, and coal. Figure 3.5 relates the resource type to the end-use fuel and engine type. Electricity can be generated from any of the fossil fuels as well as from hydro, solar, or nuclear power.

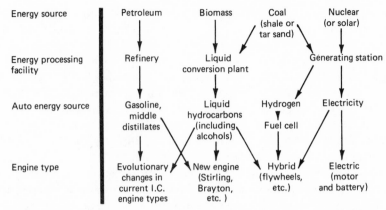

Figure 3.5   Feasible transition paths to alternative sources and/or engines for automotive transportation. (*Source*: Purdue University, Automotive Research Center.)

The most widely used engine in the transportation fleet is the gasoline-fueled internal combustion engine. That engine is the primary power plant for automobiles and trucks and the largest consumer of transport fuel. Heavy-duty trucks, buses, trains, and ships are powered predominantly by diesel engines. Commercial aircraft have turbine engines and use a distillate fuel referred to as "jet fuel." All of these engines, with some modification, can burn nearly any gaseous or liquid fossil fuel. It should be noted that synthetic gasoline, diesel, and jet fuel from coal and shale oil are virtually indistinguishable from petroleum-based fuels, and no engine changes are required for their use.

The introduction of methanol, ethanol, methane, and propane could require some changes in engine design and fuel delivery systems to accommodate different characteristics of these fuels and take advantage of some of their superior qualities as automotive fuels. For example, all of these fuels burn more cleanly than gasoline in an engine. It is likely that some of the pollution control equipment on present engines would not be necessary with these fuels. Also these fuels have a higher octane rating than gasoline, which can lead to improved performance and fuel economy if the engine is designed for the fuel.[4]

The fuel delivery systems for these new fuels will vary. Methanol and ethanol have a lower energy density than gasoline, and a slightly larger fuel tank may be required. Also, the alcohol fuels are corrosive to some materials—fuel tanks, lines, and pumps must be made with that in mind.

[4] Richard H. Shackson and H. James Leach, *Methanol as an Automotive Fuel: Report of a Workshop on Implementation Strategies and Research Needs,* Mellon Institute, Arlington, Va., November 1981.

The gaseous and cryogenic fuels require special tanks and pressure regulators, and in all cases engine carburation must be adjusted to the fuel.

Future transportation power plants such as turbines and Stirling engines are essentially multiple-fuel engines, and any liquid fuel can be used with minor modifications. Research on alternate fuels for diesels has been promising; in fact, some of these fuels may reduce diesel particulate emissions.

The cost of converting an engine to use a different fuel can be high. Conversion kits for propane or methane cost about $1200, and converting to methanol can cost up to $800. However, the extra conversion costs of high-volume, production-line engines that use any of these fuels should be negligible. With that assumption and the expected fuel efficiency gains for some of the fuels, the consumer stands to benefit if a rise in petroleum prices makes these options cost-effective.

Using electricity to power autos or trucks is hardly a new idea. When autos were first introduced, the electric car was quite popular. However, improvements in the internal combustion engine quickly made it the most desirable automotive power plant. Electric vehicles do offer a quiet, pollution-free, low-maintenance transportation option, and, depending on the fuel used for electricity generation, they can be petroleum-independent. Thus far, unfortunately, present-day electric vehicles are range-limited and thus cannot totally supplant gasoline power. But in urban areas where most of the trips are short and where air quality is in need of improvement they could someday play an important role.[5]

Some evaluation of their performance and possible role in our society is taking place. Palo Alto, California, has acquired a few electric vehicles to assess their performance in daily city functions. The Tennessee Valley Authority has a new electric-vehicle testing facility. Vehicles will be tested to evaluate distance traveled per battery charge under various driving patterns—acceleration, ability to handle different grades, and braking. Alliance, Nebraska, is using electric carts for meter-reading purposes. In all, there are about 3000 electric vehicles in service today.

Nearly all the problems associated with electric cars are related to batteries. One ton of battery is equivalent to the energy in about one-half tank of gasoline—not an effective way to store energy by comparison. Batteries wear rapidly and are expensive to replace. Thus, until battery technology improves, electric cars will remain in limited use in our vehicle fleet.

[5] H. James Leach, "Electric Vehicles: A Cost-Effectiveness Evaluation," *Fourth International Electric Vehicle Symposium,* Electric Vehicle Council, October 1981.

TELECOMMUNICATIONS

Finally, a chapter on mobility would be less than complete without some mention of telecommunications. It is believed that this rapidly growing industry of transporting information could displace some actual travel. That hypothesis has yet to be proven. In reality, the increase in telecommunications in recent years has probably stimulated travel.

Nevertheless, both mobility and telecommunication services exist in response to a demand by people to interact. The saturation point of these demands has not been established. There are many instances where the service provided is similar, and one could be substituted for the other. As the telecommunication industry develops, this relationship will be better understood, as will the trade-offs in cost and efficiency.

LEAST-COST TRANSPORTATION FUEL USE

Figure 3.6 shows the potential impact of choosing Least-Cost transportation technologies and fuels on total transportation fuel use for the next 2 decades. Even if only known technologies are designed into the fleet (this projection assumes no technological breakthroughs), total fuel use could decline by 12 percent in 2000 despite a 1 to 2 percent per year projected increase in transportation service demand. The largest potential savings are for cars and light trucks, where fuel use could drop by 40 percent over the next 20 years. Fuel use in all other modes is projected to grow, with the largest growth occurring in commerical trucks and air travel because of significant increases in projected service demand in these categories.

Automobiles are the largest fuel users by far in the transportation sector, but they also have the greatest potential for fuel savings. Fuel use in cars declines owing to relatively slow projected service demand growth (vehicle-miles traveled are projected to grow at 1.1 percent per year) and large potential fuel economy improvements. Average fuel economy for new cars (Environmental Protection Agency [EPA] value) is projected to increase from 20.9 mpg in 1980 to 38.5 mpg in 1990 and 42.5 mpg in 2000, resulting in a 47 percent decrease in projected fuel use per car. Because of a higher rate of increase in the number of light trucks, fuel use in this area will drop only 17 percent from 1980 to 2000 in the Least-Cost projection.

The heavy trucks, commercial aircraft, and miscellaneous transportation modes (buses, motorcycles, rail, marine, and pipeline fuel use) are characterized by more rapid service demand growth (3.3 percent per

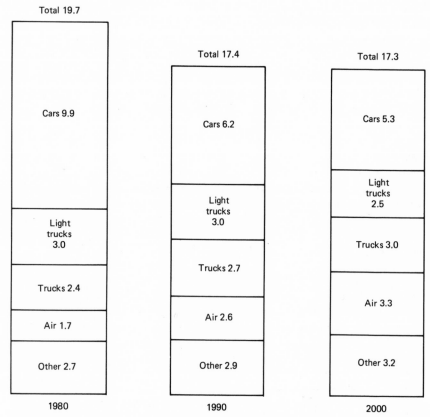

Figure 3.6 Least-cost transportation fuel use (quadrillion Btu per year). (*Source:* Applied Energy Services Least-Cost projections.)

year for heavy trucks, 4.2 percent per year for air travel, and 2.5 percent per year for the other modes) and more limited potential fuel savings (35 percent by 2000 for heavy trucks, 33 percent overall for air travel, and none assumed for the miscellaneous modes). Fuel use in these three remaining categories is expected to grow by over 40 percent, despite the projected gains in fuel efficiency.[6]

Figure 3.7 shows that when all categories and types of transportation are combined, the cost averaged $1752 per person in 1980 (in 1982 dollars). This estimate includes the cost of fuel and equipment (but not insurance, parking, or other similar costs) for all travel—including autos, trucks, airplanes, and other modes. About 52 percent was spent for fuel, and 48 percent for equipment and the carrying costs of buying that

[6] Applied Energy Services, *The Least-Cost Update,* Arlington, Va., October 1982.

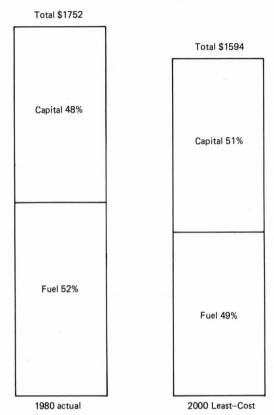

Total $1752

Total $1594

Capital 48%

Capital 51%

Fuel 52%

Fuel 49%

1980 actual   2000 Least–Cost

Figure 3.7   Mobility service costs, 1980 and 2000.
(*Source*: Applied Energy Services Least-Cost projections.)

equipment. In contrast, our projections show that if Least-Cost investments are realized, transportation costs could drop 10 percent. Capital
equipment investments and fuel costs might be about equal in the year
2000. With these investments in Least-Cost mobility options, fuel costs
could drop by about 15 percent by 2000 compared to 1980.

The evidence is powerful and convincing. As we look to the future,
abundant choices exist for ensuring mobility services at costs equivalent
to or lower than those we experience today. The data on fuel economy
options and the possibility for using synthetic fuels leads us to conclude
that the nation need not fear that it will run on empty; in fact, it could
run on full and not use a drop of imported oil, should the cost of that oil
increase substantially in the future.

4

Making Comfort and Convenience in Buildings Affordable

There is no shortage of ways to provide affordable comfort and convenience in existing or new buildings.

Comfort and convenience are two familiar and widely appreciated services people get from energy: a living or working space that is the correct temperature and humidity when the outside environment is uncomfortable, and appliances and lighting that make life convenient. Comfort might also be called "climate control"—the regulation of temperature, air movement, and humidity. In 1980 the payments Americans made for comfort in all buildings was a little under one-fifth of the total dollars spent on all energy services, or about $600 per person. The services we call "convenience" cost about $400 per person per year and include all building energy services other than space heating and cooling, such as lighting, hot water, and use of appliances.

PROGRESS TOWARD A COMFORTABLE ENVIRONMENT

The history of humanity's pursuit of a comfortable environment, especially in the last few hundred years, provides a vivid illustration of human adaptability. In prerevolutionary times, the radiated heat from wood stoves and fireplaces was the primary source of warmth for settlers. In 1742, Benjamin Franklin set forth the principle of heating a room with the warm air produced by a stove and, by doing so, vastly expanded the comfort people could obtain from a single heat source. Later systems used water as a heating medium. James Watt, inventor of the steam engine, was the first to use this principle. He heated his own office with steam in 1784 while he was exploring other applications for this form of energy. In 1801 a public building in England became the first to be heated entirely by steam. By the 1860s most commercial buildings of significance—both public and private—were heated either by steam or hot water.

In an extension of this technology, Birdsill Holly, acclaimed as the "Thomas Edison of central station heating," ran an underground pipe from a boiler in his house to his barn and later to an adjoining home. A year later, in 1877, he built the first experimental central station, or district heating plant, in Lockport, New York, which successfully heated many homes, offices, and commercial buildings the following winter.[1]

The concept of forcing hot air out into a room or building, rather than letting it find its own way, goes back at least two centuries. The word "ventilator" was invented by J. T. Desagulier to describe the person who had the job of turning the crank of the centrifugal fans that furnished air to the lower decks of ships and the English House of Commons in 1736. Even though there were some early steam and gas engine fans, it took the development of alternating current motors by Nicholas Tesla at the end of the century to make fan ventilation a feasible concept. In fact, the idea was so attractive that Edison's first electric plant in New York's Pearl Street Station had barely started generating electricity when the first use of electric fans for cooling New York hotel rooms was publicized.

In 1911, Willis Carrier was the first to provide cooling with his patented air-conditioning machine. It was initially confined to theaters, dining rooms, Pullman cars, ballrooms, and other commercial operations in which the high cost of cooling and moisture control could be offset by the increased profit gained from drawing customers who might not otherwise have patronized them. Air-conditioning did not become widely available for the home until the early 1950s, when it achieved massive popularity across most of the country as the cost of the technology and electricity decreased enough to make it affordable for most households.

Within our lifetimes these developments have led to total environ-

---

[1] Historical information on central heating from the Community Central Energy Corporation, Scranton, Pennsylvania.

mental control—the isolation of the interior of a building from the outside environment. Until recently, restaurants, retail stores, and public offices took pride in being colder in the summer and warmer in the winter than other buildings. And this was still a laudable achievement when the Arabs held back their oil for the first time in 1973.

Since that first oil-price shock in 1973 the emphasis in building-sector energy technology development has shifted from providing new services to controlling the cost of existing energy services. In the remainder of this chapter we discuss the ways in which this shift is being accomplished by all kinds of people and organizations throughout the country.

CURRENT BUILDING ENERGY SERVICES

Heating and cooling, the two components of climate control, are the largest energy services required in present-day buildings. Together they account for some 70 percent of the energy used in all buildings (see Figure 4.1). Other building services, such as lighting, hot water, refrigeration, and power for other appliances, also known as convenience

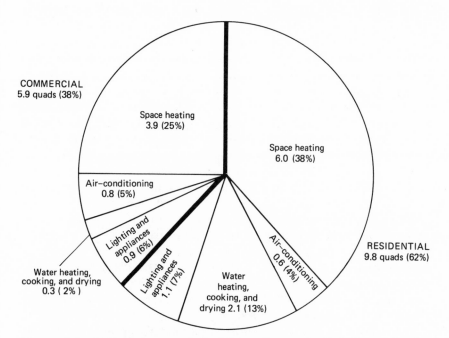

Figure 4.1 1980 energy use in buildings. (*Source*: Applied Energy Services estimates derived from Department of Energy data.)

services, make up the balance. Figure 4.1 shows that residences used over 60 percent of all the energy consumed in buildings in 1980. On average, approximately 60 percent of residential energy use was consumed for space heating. Commercial buildings have a similar pattern, but a greater fraction of their energy use is for air-conditioning. These averages vary significantly across different regions of the country, depending on the length and severity of the heating and cooling seasons in each area.

More than just energy is required to deliver building energy services. The delivery system typically includes a building envelope, or shell, which separates the conditioned inside air from the outside, a conversion device such as a furnace or air conditioner that converts energy into usable heat or cooling, and an energy source such as oil, gas, or electricity. For convenience services, the typical delivery system is simpler—usually just an energy-using device or appliance and a fuel source.

Each part of the system has an associated cost. As one component becomes more expensive, investments in other components can reduce the total cost of energy services. For example, high fuel bills have led people to increase the amount of insulation in their ceilings and walls (the building-shell component of the system). This reduces the amount of heat or cooling loss from the inside of the house and leads to a reduction in the amount of heat or air-conditioning required to keep the house at the desired comfort level. Installing more energy-efficient appliances can also be used to control energy service cost increases. Another possible method for reducing building energy service costs is to switch fuels. Thus, all three energy service components—shells, conversion devices, or fuel—offer opportunities for decreasing the cost of energy services in buildings. These improvements should not be considered separately but rather as a complete package, since one improvement may enhance, or reduce the need for, another.

Since the oil embargo of 1973 and the resulting sharp rise in energy prices, people have indeed begun to reduce their building energy service costs by spending more on insulation and other building-shell improvements as well as by installing more efficient furnaces, air conditioners, and appliances. Many energy efficiency improvements are quite mundane and easily done: for instance, the use of caulking and weather stripping to reduce the flow of unconditioned outside air into a building. Other improvements involve more ingenious design changes: for example, new homes designed to use sunlight and natural airflow patterns to provide substantial levels of comfort conditioning (so-called passive solar designs), or commercial buildings that recycle the heat generated in centralized computer rooms to provide hot water or space heat.

REDUCING THE COST OF COMFORT IN EXISTING BUILDINGS

A comprehensive study of just how much money could be saved in existing homes has been made by Robert Socolow of Princeton University's Center for Energy and Environmental Studies.[2] His research was conducted on 31 two-story townhouses constructed in 1970 at Twin Rivers, New Jersey. The work focused on space heating, since that is by far the largest component of energy used in most homes.

To any consumer worried about cost the study's findings are a source of encouragement. The center concluded that an average of 30 percent of the energy used by the townhouse furnaces could be eliminated through the routine application of relatively inexpensive measures, and savings as high as 75 percent could be achieved if special care were taken in their implementation.

Thirty-five percent of the heat lost from the Twin Rivers townhouses filtered out through the attic, despite the presence of 3 inches of attic insulation. Adding more insulation obviously helped, but airflows into the attic from other parts of the house were also found to cause large losses. By sealing cracks in the walls and "attic bypasses," such as open flue shafts, misplaced insulation, or cracks around chimneys, heat loss in the townhouses was reduced significantly.

The Princeton group also found that windows facing south can become effective solar heat collectors, while windows facing north were net losers of heat. When the researchers added a second pane of glass to the windows, they found that it tripled the net gain of heat through the windows and also helped prevent heat loss. The addition of insulating shutters on the windows, which were closed after dark, further reduced the heat that was lost to the colder outside air at night.

An additional study finding was that by caulking around windows and doors and by sealing cracks in the building structure, the number of air changes (the frequency that air within a structure is interchanged with that from outside the building) could be reduced from 0.6 to between 0.2 and 0.4 per hour. These measures led to considerable energy savings because the less the cold outside air is exchanged with the heated indoor air, the less warm air the furnace must provide to maintain a constant comfort level.

All of these measures decreased the need for furnace capacity, making it possible to heat the same-size house with a substantially smaller furnace. Though it was not carried out at Twin Rivers, when coupled with the higher furnace combustion efficiencies now commer-

[2] Robert H. Socolow (ed.), *Saving Energy in the Home: Princeton Experiments at Twin Rivers*, Ballinger Publishing, Cambridge, 1978.

cially available, furnace size reduction could further reduce home heating costs, resulting in a total decrease in heating fuel of 50 to 80 percent, while providing the same level of comfort.

A follow-up conducted by the Princeton group in 1982 demonstrated that the savings potential indicated in the first study could be replicated on a larger scale at reasonable cost.[3] In this experiment, personnel from five participating utilities were instructed as energy "house-doctors," who identified and applied low-cost measures for reducing home energy consumption during the diagnostic visit. They also identified more expensive conservation measures that could be installed during a later visit by a local contractor.

The measures most commonly implemented included caulking and weather-stripping cracks around doors, windows, and other structural features; installing set-back thermostats and gaskets around electric sockets; and adjusting furnaces to operate at maximum efficiency. Adjustments were also made to the houses' hot-water systems. These adjustments included the installation of faucet flow-restrictors, the insulation of hot-water tanks and pipes, and the lowering of the water heater temperature to 120°F if the home did not possess a dishwasher.

On the average, these measures led to natural gas savings of 15 percent, or roughly $110 per year. The cost of the house-doctor "visits" ranged from $250 to $500 per house, and the Princeton group estimated that with more practice, energy house-doctors could offer their treatments in a larger commercialized program for an average of $325 per visit.

While the house-doctor measures addressed the heat that is lost from a building because of hot-air leaks, the measures that were taken by the contractors in follow-up visits to selected study participants addressed energy losses resulting from the conduction of heat from the warm interior of the house to the colder outside. These are generally more expensive and include improvements such as adding insulation to attics, walls, and floors and installing storm windows and doors. The cost of contractor improvements in the study ranged from $650 to $2550 per house, with an average of $1250, and resulted in average total gas savings of 21 percent. Although the contractor improvements were more expensive per unit of energy saved than the house-doctor measures, the Princeton study suggests that they are still worthwhile.

These findings, similar to the results of other studies, indicate that compared to the cost of purchasing heating fuel, most of the energy-saving practices mentioned here are a bargain. A study by the National

---

[3] Gautam Dutt, Michael Lavine, Barbara Levi, and Robert Socolow, *The Modular Retrofit Experiment: Exploring the House Doctor Concept,* Center for Energy and Environmental Studies, Princeton, June 1982.

Institute of Building Sciences found that the average cost of residential conservation measures was equivalent to paying about $2.50 per million Btu for natural gas: equal to 35 to 50 cents per gallon for heating oil, or 1 to 2 cents per kilowatt-hour for electricity.[4] At current market prices these energy savings are a bargain; most residential customers are paying about twice these amounts for fuel. These findings suggest that a large supply of low-cost comfort is available by tapping the potential cost-effective energy savings in existing buildings.

Significant energy service cost savings can also be achieved in existing commercial buildings. As in residences, one of the most important factors in reducing heating costs in commercial buildings is slowing infiltration, the seeping of moist, untreated cold or hot outside air into the building interior. While blocking infiltrating air is often a good source for savings in large buildings, more important is the reduction of the huge quantities of outside air that are *intentionally* brought into buildings for ventilation purposes. One engineer estimated that in many large commercial buildings over 50 percent of the fuel used is for heating or cooling outside air pumped into the building for ventilation purposes. There are hundreds of methods for reducing the amount of outside air required for ventilation while keeping the interior air healthy and comfortable. Some examples now being adopted include installing chemical or activated-charcoal odor-absorbing devices, adding controls to shut down the ventilation system when the building is closed for an extended period, installing controls that allow "free cooling" by outside air when possible, or using stale exhaust air to preheat or prechill the incoming fresh air via an air-to-air heat exchanger.

As Table 4.1 illustrates, significant energy service cost savings are possible with investments as low as 50 cents or $1 per square foot of commercial floor space. Since the systems for providing comfort in commercial buildings are much more complex and diverse than those in residences, greater opportunities exist for making efficiency improvements in that sector, and each building must be examined individually to determine the appropriate measures for achieving cost savings. This explains the wide range of cost-per-unit savings indicated in Table 4.1.

Since the potential savings are substantial, retrofitting existing commercial buildings to reduce their heating and cooling costs has become a major business. Case studies of actual buildings show that reductions in energy in the range of 25 to 40 percent are typical. Furthermore, these investments pay for themselves very quickly: payback periods of 1 to 4 years are the norm. Because the annual energy bills of most commercial

[4] Steven C. Carhart et al., *Creating New Choices: Innovative Approaches to Cut Home Heating and Cooling Costs,* Carnegie-Mellon University Press, Pittsburgh, 1980, p. 19.

TABLE 4.1 Commercial Building Retrofit Costs and Percent Savings (Number of Sampled Buildings in Each Category)

$/square foot	% energy saved							
	0–10	11–20	21–30	31–40	41–50	51–60	61–70	71–80
0–0.50	3	4	2		1	1		
0.51–1.00	2	1					1	
1.01–1.50	1							
Total	6	5	2	0	1	1	1	0

SOURCE: Solar Energy Research Institute, *A New Prosperity: Building a Sustainable Energy Future*, Brick House Publishing, Andover, 1981, p. 158.

buildings can be quite large, energy savings of $50,000 to $500,000 per building per year are possible.

A few examples help to illustrate some of the creative measures that can be taken to reduce the cost of comfort services in existing commercial buildings. These steps do not need to involve large capital investments. The Temple Emanuel in Beverly Hills reduced its natural gas bill by 25 percent with an investment of $120. In addition to installing two time clocks to regulate the use of the temple's heating equipment, the building manager lowered the temperature to which the building was heated and also reduced the temperature of the temple's hot-water system. These activities led to natural gas savings of roughly 380,000 cubic feet, representing a cost savings of $1300 each year[5].

An improved maintenance program as well as the installation of standard building energy system controls allowed the Black Hawk County Courthouse to cut natural gas use by over 50 percent between 1980 and 1983.[6] Richard Buchanan, superintendent of the courthouse building, began his improved maintenance program in 1980. The program includes lowering the temperature of the hot water in the building, replacing worn parts in the building's steam system which lead to leaks, removing the insulating scale deposits on the heat transfer surfaces of the building's heating and cooling system, and monitoring and adjusting the efficiency of the building's boiler system for maximum effectiveness. Buchanan estimates that the cost of this program was only about $20,000 over a 3-year period.

In addition to the maintenance program, Buchanan also installed automatic dual-temperature thermostats and heat sensors in all parts of the building, which allows the heating system to cut back consumption at night. He also added an economizer device that senses the energy content of the outside air and overrides the new thermostats when

[5] "California Temple Cuts Use by 25 Percent," *Energy Users News*, Oct. 5, 1981, p. 9.
[6] Jeff Barber, "Energy Jobs Net Courthouse 1-Year Payback," *Energy Users News*, Feb. 21, 1983, p. 9.

heating or cooling can be done with outside air. The cost of the audit that determined the need for the controls and the controls themselves came to just over $37,000. Buchanan estimates that his total investment of $57,000 has yielded natural gas savings of about $50,000 per year.

In some commercial establishments, a little creative engineering is required to do a job when off-the-shelf equipment is either not available or too costly. This was the case in the Midtown Holiday Inn of Richmond, Virginia.[7] When the Holiday Inn was built several years ago, it was designed with a three-pipe system for circulating hot and cold water through the building to provide space heating and air-conditioning. At times during the year, notably the fall and spring, both heating and cooling can be required in different parts of the building at the same time. In a three-pipe system, when these streams return to the boiler and chiller room, the hot and chilled water are mixed, resulting in feed water for the chiller that is too hot and must be cooled before reuse. This practice also removes heat from the boiler feed-water stream, which must be regenerated in the system.

Gary McSherry, the engineer faced with the task of improving the system, said that upgrading the three-pipe system to a modern four-pipe system would have cost as much as $200,000 to $250,000. Instead, McSherry designed a new heating and cooling control strategy which now turns off the hotel's chiller system or boilers when the outside air is warm or cold enough to condition the building on its own. When both heating and cooling are required, the new system alternates the operation of the boiler and chiller so that the return water entering each device is at the optimal temperature. McSherry estimates that the total cost of the new control system and the replacement hot-water heater, which allowed the larger boiler system to be turned off in the summer, was $45,200, and that the entire cost of the system was recovered in natural gas and electricity savings within 9 months. More and more commercial building managers are discovering that savings like these can be achieved in their buildings, too.

LOW-COST COMFORT IN NEW BUILDINGS

The options for providing low-cost comfort in new buildings are even greater than those for retrofits of existing ones because designers do not have to work around existing structures to optimize the energy use of the system. Such options include energy-efficient combinations of conventional equipment as well as many innovative designs. For example, a

---

[7] Bob Deans, "Hotel Gets 9-Month Payback on HVAC Project," *Energy Users News,* Oct. 11, 1982, p. 4.

home in Houston owned by Andy Sansom was constructed to use no artificial energy source, taking advantage of a high sun, which does not shine directly into the house in the cooling season, and a low southern sun, which does shine into the house in the heating season. During the day the sun's energy enters the house and is stored in bricks, which release their heat inside the house at night. A large window overhang and an insulated roof protect the inside of the home from the heat of the sun's rays in the summer. Some variations of this design in southern climates use earth to provide natural insulation by burying all or a portion of the exterior walls. In all of these cases, the added cost of construction, if any, is more than offset by fuel savings, making the cost of comfort much lower than for conventional heating and air-conditioning systems.[8]

Solar-heated homes are also successful in more severe climates. For example, Larry Moss's new home in Estes Park, Colorado, has been described as an energy-conserving masterpiece and is a good example of an energy-efficient home built in an extreme climate at an elevation of 9165 feet. This house combines extraordinary insulation, high levels of solar heat gain (or the net amount of energy provided a structure by the sun), and enormous heat-storing capacity in a balanced design. The only purchased energy is electricity, which averages less than 500 kilowatt-hours per month.[9] This is far below that used in other all-electric houses. Heat is supplied to the house in three ways:

1. Solar energy moving through the building envelope or shell

2. Electrical energy coming through the meter and powering lights and appliances

3. The body heat radiated by the occupants

Innovative low-cost heating and cooling systems are also available for new commercial buildings. For example, a department store in Tuscaloosa, Alabama, included a new aquifer-based cooling system that cools water in a tower during the winter, stores it in the earth, and pumps it back through the building during summer. A comparison of the store's conventional cooling cost with the total cost of the new system (including amortization of the capital cost) reveals an expected 10 percent cost savings.[10] In another innovative application, Northwestern National

---

[8] For examples of calculations of the cost of comfort from passive heating systems see Solar Energy Research Institute, *A New Prosperity: Building a Sustainable Energy Future,* Brick House Publishing, Andover, 1981, pp. 84–88.

[9] A description of this home and its energy-using characteristics can be obtained from Larry Moss, 5769 Longs Peak Route, Estes Park, CO 80517.

[10] Bob Deans, "Store Uses 25% Less Electricity than Chain Average," *Energy User News,* July 20, 1981, p. 4.

Bank of Minneapolis reduced the cost of heating in its new operations center by 75 percent by designing a heat-recovery system into the building. This system pumps water through the company's computers, which then collects the computer's waste heat as it flows past them. In the winter this hot water is used to heat the building.[11] Again, the 75 percent cost saving was achieved even after including the amortized cost of the heat-recovery system in the calculations.

While much research and experimentation has focused on increasing the thermal integrity of building shells to reduce or even eliminate the need for an energy supply, other work has been aimed at increasing the efficiency of the conversion devices which supply heating or cooling. For example, Lennox Industries, Inc., with support from the Gas Research Institute in Chicago, has recently developed a high-efficiency "pulse combustion" gas furnace that uses approximately 25 to 30 percent less fuel than its most economical predecessor. Although the furnace is more expensive than a standard gas furnace, the fuel bill savings more than compensate for the additional cost.

Electric heat pumps,[12] which have been available since the early 1960s, represent an efficient alternative to resistance heating for electrically heated homes. As electricity prices have risen, continued research in increasing their operating efficiency and reliability has occurred, and heat pumps are now the Least-Cost choice in many applications. Natural gas–fired heat pumps, though still in the development stage and expensive, should be able to provide heat more efficiently at lower outside temperatures than the current electric heat pumps. The efficiencies of standard air conditioners have also been improved greatly and now these devices often use 75 percent less electricity than did the models developed in the 1960s.[13]

Not all novel building designs reduce the cost of services, however. For example, the University of Virginia recently installed a translucent glass-fiber roof in its new athletic facility, thus reducing electricity consumption by $34,000 per year. However, the annualized capital cost of the roof was $40,000 to $50,000 per year, thus increasing the total cost of the building $6,000 to $16,000 per year.[14]

This example reminds us that energy service costs include capital equipment and labor as well as energy and that saving energy alone does not guarantee reduced total costs. A complete cost comparison of

[11] Cindy Galvin, "Building Design Helps Bank Save on Energy Bills," *Energy User News*, Nov. 9, 1981, p. 11.

[12] A sort of refrigerator-in-reverse that taps the heat available in cold air in much the same way a refrigerator uses the cold available in warm air.

[13] Solar Energy Research Institute, *A New Prosperity: Building a Sustainable Energy Future*, Brick House Publishing, Andover, 1981, p. 77.

[14] Cindy Galvin, "University Installs $1.2 Million Roof," *Energy User News*, Oct. 12, 1981, p. 10.

different options for providing the energy services must be performed before a wise investment can be made.

Over the longer term even less expensive energy conversion systems will be introduced. Since the temperature of gas or oil burned in a furnace is in the 1400 to 2100°F range, there are many possibilities for work to be performed before the heat comes out of the register at 110°F. Electricity could be generated, water could be heated, or appliances operated with no additional fuel cost.

One energy-efficient conversion system for multifamily residences or commercial buildings is cogeneration: the simultaneous production of electricity and usable heat. For example, an office building on 42d Street in New York has recently been fitted with eight truck-type diesel engines. These engines each turn generators that supply the building's electric requirements and turn off or on automatically, depending on demand. The hot exhaust and the system's cooling water supply all the building's needs for warm air and cooling.[15] Without counting any of the benefits from a better-designed structure, this system costs 30 percent less than conventional utility-generated electricity and a steam boiler. Several other large buildings are using the same approach.[16]

In the longer run, more exotic cogeneration systems may become available for residential as well as commercial buildings. A small fuel cell that chemically converts hydrogen to electricity is currently in the advanced stages of development. The waste heat generated in the chemical reaction in the fuel cell is available for space heating and cooling in small buildings. A small free-piston Stirling engine might be able to do the same thing for a private residence in the future. While the future looks good for the development of the fuel cell and Stirling engine technologies over the next 20 years, comfort in buildings will probably be supplied by the more conventional conversion technologies and conservation options on the market today.

Figure 4.2 shows that according to our Least-Cost projections, building-shell improvements in both old and new buildings and the use of high-efficiency furnaces, heat pumps, and air conditioners could decrease the amount of fuel used for heating and cooling in buildings by 30 percent over the next 2 decades, despite a projected 15 to 20 percent increase in the building stock. For individual buildings, roughly 40 to 50 percent savings can be achieved from shell improvements alone.

Because buildings constructed before the age of higher energy costs

[15] The cooling is provided by an "absorption air conditioner," which translates the warm exhaust from the diesels to chilled air. This result could also be obtained from using gas turbines, gasoline engines, or more advanced engines such as the Stirling cycle.

[16] Pamela J. Ruben, "Firm Achieves 6 Month Payback on EMS," *Energy User News*, Oct. 12, 1981, p. 17.

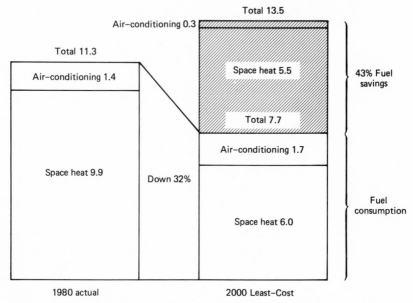

Figure 4.2   Fuel used for comfort in buildings. (*Source*: Applied Energy Services Least-Cost projections.)

will continue to be a large portion of all buildings, the largest potential building energy savings are in improvements made to existing building shells to reduce air infiltration and heat loss through wall and ceiling surfaces. By our estimates, electric heat pumps and high-efficiency gas furnaces with electric air conditioners will be the primary choices for providing comfort services in new buildings for the next 2 decades.

CONTROLLING THE COST OF CONVENIENCE

Progress in developing the efficiency and number of devices providing convenience services has also been rapid. Starting with the invention of the light bulb, people have been offered an extraordinary array of devices to make living more convenient. For the sake of simplicity, hot water, lighting, ovens, ranges, refrigerators, washers, dryers, televisions and radios, office equipment, elevators, and all miscellaneous appliances can be labeled "convenience energy services." The cost of these energy services in all buildings averages about $400 per person per year. This represents 40 percent of total building energy service costs. Only a little more than half of this is the cost of fuel.

Controlling lighting costs represents one major opportunity for savings in both residential and commercial buildings. Lighting is provided by a system of ballasts, fixtures, lenses, wires, bulbs, electricity, and natural-light reflection from the walls. In each case, the design of the lighting system depends on the intended use of the light. The system is further complicated by the heat given off by lights, which in many commercial buildings can increase the air-conditioning load significantly during most of the year.

Numerous ways exist to reduce the cost of lighting services. One method is to use light more effectively. Especially in commercial buildings, large cost reductions are possible by moving light fixtures closer to actual work areas or by moving workstations closer to available sunlight, permitting the reduction of lighting services where they are not required.

At first glance, these changes might appear detrimental to employee health. However, the reverse is actually more accurate. While in the federal government, two of us undertook a study of lighting-system energy requirements. In the course of that study, doctors in the Department of Health, Education, and Welfare were asked about the effect of lighting levels on eyestrain. To our surprise they indicated that eyestrain and related problems today are much more likely to be caused by *excessive* lighting and glare than by not enough light. With the help of these health experts, federal guidelines were established for lighting levels that were generally 50 percent, and in some cases 30 percent, of those typically found in large buildings at that time.

Lighting-cost reductions can also be achieved by replacing existing bulbs with bulbs that produce more light per watt of electricity consumed. The new fluorescent screw-in bulbs used to replace incandescent bulbs are a good example of a replacement lighting technology now available to residential users. Averaged over its normal life of 1000 hours, a typical 60-watt bulb costs about one-tenth of a cent per hour. At electric prices of 5 cents per kilowatt on average, the electricity cost of lighting the bulb is about nine-tenths of a cent per hour. Thus 90 percent of the cost of light from incandescent bulbs is the cost of electric current. For the replacement fluorescent bulb, offered for the home by General Electric and others, the cost per hour of a 60-watt bulb equivalent is two-tenths of a cent per hour, twice as much as the standard bulb. On the other hand, the electric cost is only five-tenths of a cent per hour, almost half the operating cost of the incandescent bulb. Thus, the total cost of light services using the new bulb compared to the old is 40 percent less while requiring roughly half as much electricity. Incandescent bulb improvements have also been made for both standard and three-way bulbs. Assuming that a combination of these efficient bulbs is used in the future, the total electricity demand for providing lighting services in residences could remain roughly constant over the next 20 years, including the need to light the roughly 25 percent increase anticipated in residential housing units.

For commercial and industrial lighting situations, General Electric has introduced the Optimiser fluorescent lamp and ballast that uses 33 percent less electricity compared to a standard 40-watt fluorescent fixture (two bulbs and ballast) with a similar light output. At the published price of $3 per bulb (1982 dollars), the capital cost per hour over the bulb's projected 15,000-hour life is one-tenth of a cent per fixture, about the same as a standard fluorescent lighting system. However, the electricity cost of the new system is 33 percent less than for the standard fluorescent fixture. The chart in Figure 4.3 illustrates the total cost of this new, efficient General Electric Optimiser fixture compared to a standard fluorescent lighting system. The left column shows the elements of lighting costs, using the old bulb. Adding the new bulbs keeps the fixture costs about the same and lowers electricity consumption. The net result is roughly a 30 percent reduction in the total cost of the lighting services.

Since General Electric is particularly effective in marketing new products, the more efficient light bulbs will probably sell. But wouldn't it

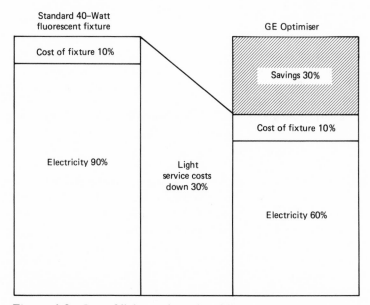

Figure 4.3 Cost of light can be reduced by new bulb technologies. (*Source*: Operating data from the General Electric Company.)

be preferable if that development were not a "disadvantage" to the electric utility, as Edison had put it? Under the current system, if everyone buys the new bulbs, the utility's future electric sales for satisfying lighting needs will be much less than what they are today. If Edison had his way and utilities were still selling light-hours instead of kilowatt-hours, electric utilities would still be providing total lighting services (instead of just electricity), only at a more favorable and competitive price.

Similar kinds of opportunities are being developed to reduce the costs of providing hot-water services for homes. By now, almost everyone has either seen or installed a water heater insulation blanket. The savings for this activity usually run in the range of 15 to 20 percent. Consequently, most new water heaters are constructed now with greater amounts of insulation than they had 10 years ago. In some climates, solar water heaters are able to provide hot water at costs much lower than electric or even gas-fueled systems. The economic attractiveness of solar hot-water systems may increase in future applications as the technology matures, collector costs decrease, and the cost of alternate fuel choices increase. Even more efficient heat-pump water heaters that appear to save 50 percent on the average over conventional electric water heaters are just beginning to be marketed by a few companies.[17] Some people are using another method to lower water heating costs; by installing water pipes in the flue of their furnaces they can use the remaining energy in the furnace exhaust to heat or preheat the water before it goes to the water heater.

The use of so-called waste heat to preheat hot water has also been used in commercial buildings. For example, when Gary McSherry revised the heating system of the Midtown Holiday Inn, he also installed a heat exchanger for $3500 which removes the heat exhausted by the building's refrigerators and uses it to preheat the water used in the hotel laundry facility.[18] Other commercial buildings are using the heat generated by computers and other devices to the same end.

At the other end of the faucet, reducing the consumption of hot water through the use of flow-limiting shower heads, for example, is another important and often overlooked way of reducing the cost of hot-water services. All of these measures will be used to reduce the demand for energy needed to supply hot water in buildings over the next 10 or 20 years.

[17] For information on heat-pump water heaters, see E-Tech, Inc. and Energy Utilization Systems marketing brochures; also, U.S. Department of Energy, Office of Building and Community Systems, *Demonstration of a Heat-Pump Water Heater*, vol. 2 (ORNL/Sub-7321/4: June 1981).

[18] Bob Deans, op. cit., p. 4.

In addition to home owners or commercial building managers, other parties are also getting involved in the effort to lower the cost of energy services in buildings. In some areas, electric utilities are proposing to replace all residential electric water heaters in their service area free of charge. In return, they will install a replacement electric-powered system that only operates in off-peak hours. Highly insulated holding tanks maintain the desired water temperature at other times. While this device doesn't actually save electricity, it utilizes the capacity of an electric generating plant more evenly during the day, thus reducing the cost to the utility of generating the electric service.

Spreading electricity use away from peak periods allows a utility to lower its expenses for installing new peak-load generating capacity. Off-peak electricity costs are thus much lower than peak-period costs because they don't have to include the cost of depreciating the more expensive peaking power plant units, which are operated for only a fraction of the year. (That cost is fully borne by the peak-period users.) Thus, off-peak power costs can be as much as one-third to one-fourth the cost of peak power because the cost of the base-load plants, which generate off-peak power, can be spread over a higher level of electricity sales.

A recent development in the control of energy services in buildings are manual or computer-controlled devices known as energy management systems. These devices measure and regulate the use of energy in buildings for such diverse activities as lighting, air-conditioning, and escalator or elevator operation.

Most energy management systems are also designed to take advantage of the cost differential between peak and off-peak power. For many commercial users, utilities charge not only for the amount of electricity used each month but also for the maximum amount used at one period during the month. This rate structure provides incentive for businesses to level the peaks in their energy consumption. Energy management systems achieve this by shutting off nonessential equipment during times of high electrical use. The system may also control thermostat settings in various areas to reduce space-conditioning energy demands when all or part of the building is not in use. In a few homes microprocessors also match heating and cooling needs to those parts of the house most frequently used at a given time.

Most users of energy management systems report significant savings from the devices; some are even pleasantly surprised at how well they perform. When an energy management system was installed at Wesleyan University in the fall of 1981, annual fuel and electricity savings of $100,000 a year were anticipated. However, as the system was fine-tuned to turn off air-conditioning in unoccupied rooms and prevent oversized air-handling equipment from operating more than necessary, there were additional savings of $48,000 in steam costs, and $47,000 in electricity costs. These savings indicate that the $350,000 system should pay for itself in slightly less than 2 years.[19]

Other convenience services have also witnessed considerable cost-saving advances. Cost-reducing improvements in energy efficiency have taken place for all major home appliances. In the case of refrigerators, current models consume roughly 45 percent less electricity than those produced in 1972 and cost only 3 percent more to purchase.[20] Continued energy reductions are expected from increased insulation and new designs. For other appliances, such as televisions and stoves, technological advances have made the units more energy-efficient without adding to manufacturing or purchase costs at all. Improvements are expected to continue for most household appliances, reducing the energy services cost and energy consumption still further.

In addition to lighting and hot-water heating, the cost of other convenience services in commercial buildings can be reduced. In these structures, the cost of internal transportation provided by elevators can be quite high and often offer major opportunities for cost reductions. While the elevators in many buildings operate approximately 14 hours a day, 6 days a week, elevator capacity is designed for peak periods of use. By design, then, there are many off-peak hours when elevators are greatly underutilized. Such was the case in a 1-million-square-foot office building in Manhattan we worked on. In this building, a new procedure was adopted to take several elevators out of service during off-peak times. In establishing the schedule, care was taken to ensure that the time tenants had to wait for elevators was acceptable. The new schedule reduced elevator-operating time, and consequently electricity use dropped by 25 percent, or $40,000 per year.[21] But an even more important factor in calculating the effect on the total cost of service was the resulting extension in the operating life of the equipment and reduced maintenance, which resulted in a net reduction in elevator service costs of over $100,000 annually.

The elevator illustration underscores common problems that often exist in both buildings and industrial processes. Most systems that are built to provide energy services are designed to meet peak-load conditions. Often, however, services are not needed at that maximum level, either because of the time of day and the manner in which tenants use

[19] Lisa Cohn, "EMS Lower Fuel Costs Save University $250,000," *Energy User News*, Feb. 23, 1983, p. 14.

[20] Solar Energy Research Institute, *A New Prosperity: Building a Sustainable Energy Future*, Brick House Publishing Co., Andover, 1981, pp. 74, 75.

[21] Authors' personal experience.

the building (i.e., not occupied on nights and weekends) or because of external factors like humidity, the brightness of the sun, and temperature. Reducing the cost of energy services in existing buildings can be achieved in large part by installing equipment and operating procedures that ensure that the level and quality delivered matches the services required by the building user.

THE AGGREGATE ENERGY SAVINGS WILL BE LARGE

Our Least-Cost projection of energy use in both residential and commercial buildings is shown in Figure 4-4. As a consequence of implementing cost-effective measures that reduce energy consumption in existing and new buildings, total energy use in the residential and commercial sector could decline over 25 percent by 1990, even though the number of housing units is projected to increase by 14 percent and the amount of commercial space is projected to expand 21 percent. By 1990 most of the retrofit activity might be completed, and gradual growth in energy use (coupled with growth in the building stock) will resume. However, because the new buildings will be much more efficient than existing ones, total energy use should grow less rapidly, at less than 1 percent per year after 1990.

Most of the energy used in buildings between now and the end of the century will be consumed in buildings that already exist. Consequently, the most important feature of the Least-Cost strategy is the introduction

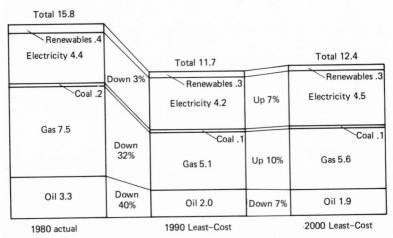

Figure 4.4 Buildings fuel use (quads per year).

of energy productivity improvements in existing residential and commercial buildings. Improvements in the thermal integrity of existing buildings could reduce their heating-fuel consumption by 45 percent and air-conditioning energy requirements by 25 percent. Total energy use per building might be reduced by almost 40 percent on average, with the bulk of these savings coming from reductions in space-heating requirements.

Furthermore, changes in the Least-Cost mix of fuels used in buildings are projected to be significant. If oil prices rise in real terms over the long run, as we expect, oil use should continue to decline in buildings. Following the trends of the late 1970s, existing oil-heated units will be retired or replaced by units heated with other fuels, and fuel use in the remaining oil-heated homes will be reduced by the installation of building-shell improvements such as increased insulation and storm windows. Our results also indicate that gas use may decline through the 1980s, again owing to conservation measures, but then resume its slow growth in the 1990s as it is selected as the Least-Cost fuel choice in many installations. In some regions, gas provides the Least-Cost space-heating option, and, where available, gas water heaters are always the cheapest option.

Electricity demand in buildings will likely remain fairly constant over the next two decades. Our projected electricity demands reflect the net effect of retrofitting electric-resistance-heated residences, increased use of heat pumps and highly energy-efficient construction practices in new

TABLE 4.2 Per Capita Cost of Energy Services in Buildings (1982 dollars)

	1980	2000	% change
Residential			
Comfort	359	358	—
Thermal*	130	139	+7
Lighting and appliances	172	161	−7
Subtotal	661	658	—
Commercial			
Comfort	255	233	−9
Thermal*	12	11	−8
Lighting and appliances	98	90	−8
Subtotal	365	334	−9
Total	1026	992	−3

* Thermal includes water heating, cooking, drying, washing clothes and dishes in residential buildings, and water heating in commercial buildings.
SOURCE: Applied Energy Services, Inc., data and Least-Cost projections.

buildings, and the installation of more efficient lighting and appliances—particularly refrigerators. Renewable energy sources, especially solar water heating and space heating and wood space heating, should account for relatively small fractions of total energy use in this period. It must be noted, however, that the cost-effective passive solar heating practices discussed earlier are considered building improvements rather than energy consumption, so our renewable energy totals do not measure this contribution.

Because more efficient energy technologies are expected to reduce fuel consumption and provide energy services at a lower cost, the total cost of buildings energy services per person should drop over the next two decades as a Least-Cost strategy is followed. Table 4.2 shows that building consumers can offset all of the expected increase in fuel prices with energy efficiency improvements—with energy services costing 3 percent less in total by the year 2000.

NEW WAYS TO MARKET BUILDING ENERGY SERVICES

Perhaps the most interesting developments in building energy improvements focus not on technology or engineering improvements but on changes in the marketing of building energy services. One of the major problems in fully implementing all of the efficient devices and conservation techniques mentioned in this chapter is the multitude of options. The choices are so varied and complex that a qualified engineer is required to design a Least-Cost system. Yet for the most part, such design services have not been conveniently offered to consumers. Also, commercial building owners are primarily in the business of renting floor space and for the most part have been able to pass through to their renters all energy service costs. With rental space fairly tight, they have little incentive to reduce the cost of energy services to their tenants. Recently, however, as occupants have focused more attention on energy, several new programs have been introduced which help the building owner or developer lower operating costs.

For example, Pacific Power and Light provides a complete home retrofit package for customers in Oregon, Washington, Montana, and Idaho. PP&L first conducts a free home-energy audit and then determines a set of "economic" retrofit measures. Customers are informed of the utility's recommendation and asked for permission to retrofit. PP&L selects the contractor, oversees the work, conducts postinstallation inspections, and pays the contractor, creating a loan on the customer's account. Customers are charged no interest and must repay the loan only at the time the house is sold. PP&L adds the amounts loaned out to the capital base on which they are allowed income from customer's rates but subtracts them as loans are repaid. Results from the program are encouraging: almost 60 percent of the owners who have an energy audit done on their home agree to go ahead with the full program.

The Tennessee Valley Authority has a program similar to that of PP&L. In cooperation with its 160 power distributors, it offers a free audit to both its heating and nonheating customers, followed by an interest-free loan to finance recommended retrofits. Landlords and renters as well as home owners are eligible. Unlike the PP&L program, the loan is payable in 7 years. Participation in the TVA program is running over 40 percent with retrofit investments averaging $310.[22]

In a more dramatic effort, General Public Utilities has proposed a program for their customers in New Jersey and Pennsylvania in which the utility would pay for all the costs of making selected cost-effective energy efficiency improvements. Under the GPU program, the customer would have no loan to pay back. The theory behind this is that the selected improvements would cost the utility less than building new power plants, thus making it a good deal for everyone. Approval of this proposal would make GPU a complete energy service company, the first utility of its kind to heed Edison's advice and expand its business to delivering energy services, not just electricity.

As a variation on the GPU theme some utilities are establishing new subsidiary companies to market building energy services. For example, Central Hudson Enterprises Corporation, a subsidiary of Central Hudson Gas and Electric Corporation (CHG&E) in Poughkeepsie, New York, markets a "shared savings" plan to commercial establishments. The company will finance the consulting work for energy conservation projects in return for a share of the energy savings (if no savings occur, the customer owes nothing). One CHG&E customer saved $70,000 per year because of this program, with 20 percent of these savings going to Central Hudson Enterprises.[23]

Other nonutility businesses are also getting into the energy services market. For example, a subsidiary of Royal Dutch Shell has an energy service company called Scallop Thermal Management Corporation, which sells heat and hot water to multifamily residences and commercial buildings at a fixed contract price that is calculated to be lower than the building's anticipated energy costs. In this arrangement Scallop is responsible for operating the heating plant (furnace or boiler) and the

---

[22] Steven C. Carhart et al., *Creating New Choices: Innovative Approaches to Cut Home Heating and Cooling Costs*, Carnegie-Mellon University Press, Pittsburgh, 1980, pp. 34, 35.

[23] Lisa Cohn, "Shared Savings Nets Center 46K in 9 Months," *Energy User News*, Mar. 21, 1983, p. 5.

cooling and heating system controls as well as maintaining or improving the integrity of the structure, all in an integrated system. Potential energy savings are generous enough that Scallop customers are paying approximately 15 percent less than they did before Scallop arrived, and they have no need to worry about the severity of the winter, since Scallop pays their fuel bills.

Another energy service program, started in New Jersey, seeks to stimulate cooperative ventures between private energy service companies and investor-owned public utilities. In the fall of 1981, an energy services company, National Conservation Corporation began systematically auditing large, contiguous blocks of homes and hiring local suppliers and contractors to offer to carry out recommended retrofits for audited customers free of charge (contractor fees are instead covered by NCC). In this first experiment of 1000 homes, NCC is being paid for the value of saved energy by the electric or natural gas utility, which will avoid the cost of buying natural gas or building additional electric generating capacity to service the demand that is displaced by the home improvements. A measurement system will insure that the utility will pay NCC only for the value of the energy that is actually saved. The cost paid by the utility will be somewhat less than the utility would have to pay for new, traditionally produced electricity or natural gas, so that other utility customers will also benefit from the improvements. The utility's costs for new capacity will be determined by the New Jersey Public Utility Commission, and the amounts paid for "saved" energy will be spread to all customers.

In a grass-roots energy service program, citizens in Rhode Island have established a cooperative effort comprised of private corporations, state and local banks, and contractors. Rhode Islanders Saving Energy (RISE) seeks to assist all Rhode Island property owners in reducing their energy bills. To do this RISE (1) conducts free energy audits, (2) selects approved contractors, (3) supervises and inspects contractor work, and (4) offers reduced-interest loans from local banks and audit companies.

RISE's cost of doing business is about 25 percent of the total amount, which is passed along to their customers. However, the total price is still competitive because RISE pays contractors a wholesale price and customers are charged lower interest rates than they would have been able to obtain individually.

In another local effort, the people at Fitchburg, Massachusetts, rekindled their pioneer spirit by being one of the first communities to organize energy conservation measures for their town. Fundamental Action to Conserve Energy (FACE) created a poster that told the story: "Send the Ayatollah a message—Fitchburg doesn't need his oil."

Scores of volunteers from the town showed up for training programs and the town went into action; armed with caulking guns, they sealed leaky homes, used weatherstripping, and took other low-cost steps to retrofit their homes. Household energy bills were cut by an average of 20 percent. Sixty percent of Fitchburg participated in this effort, and most used the more obvious, cheapest and easiest-to-install methods to reduce their energy bills. Follow-up studies on the Fitchburg experiment later revealed that more information and training was needed to educate residents on additional, more complicated improvements they could make and that lack of funds had prevented them from making the more complete changes that could have saved them even more money.

But this is just the beginning. We anticipate that in a market where many commercial organizations are competitively offering energy services and guaranteeing the results, the need to educate millions of people about the intricacies of their energy systems will be eliminated. What is abundantly clear about all these examples and analyses is that there is no shortage of ways to provide comfort and convenience in existing or new buildings economically, and the cost of comfort need not be any higher by the end of the century than it is now. Increasingly, people are recognizing that their much-discussed energy demand is really a demand for comfort and convenience, not fuel. And as people demand lower-cost services, organizations are responding with creative and effective programs to provide them.

5

Lowering the Energy Cost in Industrial Products

Energy productivity is providing a powerful means to improve product costs, rather than an option of "last resort."

While the progress in mobility and buildings is more familiar and understandable, the actions being taken in industry started earlier and produced more immediate results. As with consumers in general, the sudden increase in energy prices and the energy supply disruptions during the 1970s created a new set of headaches for managers of energy-using companies. For decades before 1973 industrial energy prices had remained stable or even declined, and their businesses prospered. Between 1973 and 1981 they saw the price of industrial natural gas rise at an average of 28 percent per year, while distillate and residual fuel oil prices rose 26 percent per year, and the prices of electricity and coal, 14 and 11 percent, respectively.[1] Since these increases were much greater than the prices of other goods and services, the total cost of purchased fuels and electricity as a percent of the value of manufactured goods also increased dramatically—almost 100 percent in the five most energy-intensive industries (see Table 5.1). Although these statistics seem depressing, today's managers are continuing to fight back with a whole variety of alternatives for reducing fuel costs in their businesses.

[1] Bureau of Labor Statistics, Producer Prices and Price Indexes.

73

TABLE 5.1   Energy Costs of Selected Industries

Industry	Cost of purchased fuels and electricity as a percent of manufacturing value added		
	1973	1980	% increase
Primary metals	10	20	100
Petroleum refining	10	16	60
Stone, clay, and glass	8	15	88
Paper	7	16	129
Chemicals	6	12	100

SOURCE: Annual Survey of Manufacturers for 1973 and 1980.

IN THE BEGINNING: CONSERVATION AS SACRIFICE

The shortages and higher prices that arrived with the oil embargo of 1973 created a great appreciation among industrial managers for the benefits that flow from energy conservation. As originally perceived in those days, conservation meant doing without. Motivated to save money, a number of managers curtailed some of their industrial energy use by turning down thermostats, buying smaller and less comfortable cars, driving trucks at slower speeds, and forming car pools for employees, to name just a few examples. However, simply doing without carried with it the burden of reduced employee satisfaction and production cutbacks. Unfortunately, sacrifice was the major form of conservation being promoted, so it was not surprising that many executives feared that mandatory conservation was around the corner and they condemned it as an anti-American trend.

At the same time conservation was also discounted as a serious solution to the energy crisis. In spite of the fact that industrial energy productivity had been improving 1 to 1.5 percent per year before the embargo, the Chase Manhattan Bank stated in 1972: "An analysis of the uses of energy reveals little scope for major reductions without harm to the nation's economy and its standard of living. . . . There are some minor uses of energy that could be regarded as strictly non-essential— but their elimination would not permit any significant savings."[2]

EFFICIENCY INSTEAD OF SACRIFICE

Fortunately, a different view of energy conservation soon began to take root in American industry. More efficient use of energy became a major part of every industry's strategic plans to reduce costs through improved

[2] John G. Winger et al., *Outlook for Energy in the United States to 1985*, Chase Manhattan Bank, New York, 1972, p. 52.

management of industrial systems. Results in manufacturing were striking: companies discovered that better energy management could often reduce energy requirements by 10 to 30 percent without major capital investment or reduced production.

In fact, the record reveals that since the oil embargo, improved energy productivity in the industrial sector has significantly exceeded most expectations. For the period 1973 through 1982, industrial output of goods and services rose 7 percent, while industrial use of energy decreased 19 percent—roughly a 24 percent overall improvement in energy productivity, or an efficiency increase of about 2 percent per year.[3]

Significant efficiency improvements have been achieved in all industries. According to the Department of Energy, the five most energy-intensive industries saved an average of 20 percent of energy consumed per unit of output from 1972 to 1981, while the next five averaged 26 percent energy savings (see Table 5.2). The indication is that gains in productivity were somewhat easier to achieve in the less energy-intensive industries, possibly because energy-intensive industries have had greater incentives in the past to operate more efficiently and minimize their higher energy costs.

TABLE 5.2 Industrial Energy Savings, 1972–1981

Energy-intensive industries	% energy savings*
Petroleum and coal products	21%
Chemicals and allied products	24%
Primary metals	14%
Paper and allied products	23%
Stone, clay, and glass products	21%
Average of top 5 energy users	20%
Food and kindred products	22%
Fabricated metal products	23%
Transportation equipment	32%
Nonelectric machinery	29%
Textile mill products	19%
Average next 5 energy users	26%
Overall average	21%

* Reduction in energy use per unit of industrial output (measured by value added in manufacturing).
SOURCE: U.S. Department of Energy, *The Industrial Energy Efficiency Program*, Washington, D.C., 1981, pp. 4–5.

[3] Industrial energy data from Energy Information Administration, *Monthly Energy Review*, Aug. 1983; data on industrial production from the U.S. Department of Commerce, *Survey of Current Business*, vol. 62, no. 4, April 1982, p. 20.

Most of the energy efficiency improvements that have occurred are attributable to economic incentives provided by higher energy prices. Price increases have caused both manufacturing improvements and a significant change in the mix of industrial products required by consumers. Of course, changes in product mix occur as Americans alter their lifestyles and spend money on different goods and services. But rising energy prices have almost certainly influenced some of these changes in product mix. For example, if product A consumes more energy in its production than does product B, the price of product A will increase faster than the price of product B, all else remaining constant. If the two products are easily interchangeable, then as energy prices increase people will consume less of product A and more of product B. This shift alone should cause a decrease in industrial energy, as has been the case over the last decade. By one analyst's reckoning, over one-quarter of the reported industrial energy efficiency improvements were caused by shifts to the production of less energy-intensive products.[4]

The large industrial energy productivity improvements of the past decade suggest that the conventional wisdom of only a few years ago has been turned on its head. These improvements have lowered the costs of industrial energy services, as industrial managers fought back against high energy prices. The results to date are better than the government or industrial representatives expected at the time of the oil embargo. Despite the recent decrease in energy prices after 1980, energy managers continued to make productivity improvements.

HOW DID THEY DO IT?

In the Mellon Institute work we discovered several factors which have contributed to the success of industrial energy efficiency improvement programs. One of the most important of these is a basic shift in the way that management views the use of energy. Only a decade or so ago energy managers were as unknown as the energy problem. Energy expenses were viewed as an insignificant part of total manufacturing costs. It was common practice to lump all energy costs into a general or manufacturing overhead account without paying them notice.

The reality of high and quickly escalating energy costs brought major changes to this viewpoint. Today almost every corporation employs energy managers, from the corporate to the plant level, and it is being proven in corporate circles that investment in conservation or energy

[4] R. Marlay, *Industrial Energy Productivity, 1954–1980*, Massachusetts Institute of Technology, Cambridge, 1981.

productivity is smart business. As Dick Aspenson, the energy director of 3M explained, "Energy costs are escalating at a higher rate than other costs of doing business. . . . For that reason, projects that cut energy costs can improve a company's bottom line more than projects that trim other costs." He noted further: "If we save $50 million in one year in energy costs, as we did in 1981, it would take $250 million worth of Scotch tape sales to generate that same profit."[5]

Successful managers of energy efficiency improvement programs focus on the total cost of the specific function or service provided by the use of energy, not just on the amount of energy consumed.

As with energy generally, industrial energy in its primary forms, such as oil, natural gas, coal, solar energy, and uranium, and in its carrier forms, such as electricity and steam, should be distinguished from the industrial energy services that it provides. Examples of these services are the heat treating of metals, petroleum and chemical distillation and evaporation, paint and fabric dying, or machine drive to power fans, pumps, and conveyor belts. In recent years, it has become evident that many of these industrial energy services, like those in the buildings and transportation sectors, can be delivered with less raw energy by substituting more energy-efficient technologies or processes for older, less-efficient systems. Such technologies are abundant, and higher energy prices and the prospect of shortages have provided the necessary incentives for their commercialization and utilization.

The existing data about industrial energy services are imperfect, but they do indicate the benefit of this approach. For instance, Table 5.3 shows that the aggregate industrial expenditure for energy services in 1980 was $152 billion (in 1982 dollars); 23 percent of that was spent on energy for machine drive from electric motors, 25 percent for the raw chemical and petroleum feedstocks that are processed into end products, 48 percent for steam and process heat, and 4 percent for electrolytic functions. This focus on the total cost of industrial energy services yields a very different picture than if we merely measure these functions in energy units. For example, the 23 percent of the dollars expended for machine drive provided only 14 percent of the energy services. Feedstocks, in contrast, required only 15 percent of the dollars but provided 32 percent of the energy services.

In most organizations management is much more specific about the energy services provided than the categories in Table 5.3. And some even measure the *total* cost, not just the energy cost. For instance, a midwest chemical plant routinely measures the total cost of its energy

[5] "Energy Manager Stresses Need to Sell Conservation," *Energy User News*, Aug. 16, 1982, p. 8.

TABLE 5.3 Annual Cost of Industrial Services in 1980 (billions 1982 dollars)

Services	Annualized capital	Fuel cost	Operation and main-tenance	Total	Energy services (quads per year)	Average cost per Btu
Energy services						
Machine drive	1.8	31.1	1.9	34.8	2.5	13.92
Process steam	7.0	20.3	7.3	34.6	4.6	7.52
Indirect heat	1.1	15.6	1.2	17.9	1.9	9.42
Direct heat	0.9	13.0	0.9	14.8	0.8	18.50
Electrolysis	0.1	6.5	0.1	6.7	0.5	13.40
Other process heat	0.6	4.7	0.7	6.0	0.2	30.00
Subtotal	11.5	91.2	12.1	114.8	10.5	10.93
Feedstocks	—	22.8	—	22.8	5.9	3.86
Other*	—	14.7	—	14.7	2.1	7.00
Total	11.5	128.7	12.1	152.3	18.5	8.23

* Primarily road oil and asphalt.
SOURCE: Applied Energy Services, Inc. industrial demand data.

services—including fuel costs, depreciation, maintenance, and labor. The analysis allowed frequent trade-offs to be made between labor, capital, and energy to minimize costs. This enabled the company to control the total cost of each energy service, not just the fuel costs, which led to substantial energy savings.

Another element of successful industrial energy programs is to control energy functions as a product cost, not just as a part of total manufacturing costs or general overhead. As mentioned previously, it was common practice in industry to combine all energy costs into one general or manufacturing overhead account without identifying those products with the highest energy service cost. The experience of a large chemical plant illustrates the benefits of product cost orientation. After the energy service costs had been carefully broken out, the company found that its previous gross margin calculations (which excluded energy) were erroneous and completely reoriented its marketing efforts to emphasize different and more profitable products.

In one other case, the starting point for reducing energy costs was to estimate the minimum cost possible for each energy service with present equipment and processes. The company then monitored its actual expense and took action to reduce the difference. It was only at this point that a change in process or equipment configuration was considered. The company recognized that an equipment change prior to taking steps to minimize the expenditure under the present system would lead to the purchase of too much new equipment or perhaps even the erroneous replacement of equipment for functions made unnecessary by the first overhaul.

EMPHASIS ON CONTROLS

Some energy management programs have also achieved success by installing controls and procedures that monitor energy costs on a regular basis. While it is commonplace to gather general knowledge about how large amounts of energy can be saved in a plant, it is unique to develop the discipline necessary to achieve these potential savings. This usually requires frequent monitoring of each service by the first-line supervisor, who is trained to see noticeable changes. At Union Carbide's Texas City, Texas, plant, for example, costs are monitored daily; fuel usage deviations are constantly logged and corrective actions taken. At this plant, the energy manager has the authority to take appropriate action, rather than merely advising management.

Even with all the progress to date, the potential for further increasing energy productivity and lowering energy service costs remains substantial (24 percent since 1973). Recent gains in industrial energy productivity appear to be just the beginning. To quote Ron Wishart, the energy director of Union Carbide, "Long-term results will be much greater."

ENERGY MANAGEMENT

A multitude of technologies are available for managers to use in minimizing the cost of energy services in their industrial plants. But the lowest cost and most easily implemented energy steps are generally referred to as energy management or "housekeeping" practices. There is no one strict definition of what housekeeping measures are, but in general they involve minimal capital investments (often written off as expenses in the year in which they were made) and often involve changes in plant operating or maintenance practices. While these housekeeping measures are frequently as mundane as their name implies, they can lead to substantial energy service cost savings. In fact, industry analysts have estimated that energy management or housekeeping activities could have accounted for roughly half of the industrial energy savings occurring since 1976 in the four major energy-consuming industries. Potential future housekeeping savings are estimated to be as large as 10 percent of 1980 industrial energy use.[6]

Energy management activities can be applied to all aspects of a plant's operation, from steam generation and use to furnace and motor operation, plant lighting systems, and plant space conditioning. For example, a poultry processing plant in California found that keeping the door to the boiler room closed raised the temperature of the air entering the boiler by 17 degrees and saved the company over $7000 per year.[7] An automobile manufacturer found that it could substitute lukewarm or cold water for heated washes in its parts-washing facility. The decreased steam requirements resulting from this move saved the company over $123,000 per year and required only a minimal cost for new detergent chemicals. These examples are not unusual; most plants have provided similar opportunities for lowering energy service costs by modifying wasteful operating practices that were developed in the days of cheap energy.

Although space-conditioning systems in industry do not require as large a percent of the total energy budget as they do in commercial buildings, heating, ventilation, and air-conditioning (HVAC) systems can offer a plant manager substantial energy-saving opportunities. One way of lowering plant space-conditioning costs is through the use of economizer controls on HVAC equipment. These controls operate dampers that allow a building to be heated or cooled by outside air when appropriate, reducing the load on the internal heating or cooling system. For example, a graphics design plant in California found that the use of economizer vents in its HVAC system led to energy savings of 365,000 kilowatt-hours per year, or roughly $23,000.

Other system changes can lower HVAC costs. Twenty-two Allis-Chalmers plants installed electrostatic and mechanical cleaning filters in their HVAC system to remove dirt and fumes from the air delivered to the work area. Before that time, outside air, which required heating and cooling, had been used to dilute the dirty plant air. Many of the plants realized HVAC cost savings on the order of 40 percent after implementing the filter systems.

The installation of simple timing devices is another way to decrease space-conditioning costs. The Dennison Manufacturing Company of

[6] Energy and Environmental Analysis, Inc., *Industrial Energy Productivity Project (Final Report): Volume 8*, prepared for Mellon Institute's Energy Productivity Center, April 1982, pp. 120–121.

[7] Many of these examples of industrial energy efficiency improvements were taken from Deborah J. Long, *Industrial Energy Productivity Handbook*, compiled for the Mellon Institute Energy Productivity Center, December, 1981, under a grant from the U.S. Department of Energy.

Massachusetts found that annual air-conditioning costs for their paper plant were cut by $7000 when they installed simple 7-day clock timers to control the plant's 13 air conditioners. Again, these savings are not exceptional; most plants have dramatically cut their space-conditioning bills through measures like these and by taking additional steps like installing loading-bay doors or storm windows to minimize the heat loss from the plant structure.

The commonplace incandescent light bulb, although familiar, is unfortunately not a very efficient lighting source. In situations where daylight quality lighting is required, fluorescent lighting systems can provide over three times the light output of standard incandescent lighting systems for the same energy input. Lighting systems with even higher efficiencies are also available: metal halide lamps can provide up to 9 times the light output of an incandescent lamp of equal wattage, and high pressure sodium lamps can provide as much as 10 times the light output of an incandescent lamp using the same energy input.

However, these lighting systems have characteristics which are not suitable for all applications (high-pressure sodium lights have a graying effect on reds and blues, for example). The energy manager designing a lighting efficiency improvement program must therefore match the work situation with the lighting source, providing the light quality required at the lowest energy input.

Not all lighting efficiency techniques require the installation of new lighting systems; simply cleaning existing light fixtures on a regular basis can allow a lighting system to deliver 10 percent more light output at the same electricity input. Also, lighting levels in many businesses are much higher than necessary, and lamps can be replaced with lower wattage lamps or removed altogether with no detrimental effects on plant production.

As an example of the savings possible from a lighting efficiency program, the Allis-Chalmers plant in Matteson, Illinois, undertook a program to:

- Reduce lighting in nonwork areas
- Clean and replace degraded light fixtures
- Install efficient fluorescent lamps and reduce the number of lamps in most fixtures from four to two
- Replace mercury vapor lighting with high-efficiency, high-pressure sodium lighting.

The electricity cost savings from this program was estimated at over $26,000 per year. Other companies have achieved similar results. The

National Can Corporation replaced its standard mercury vapor lamps with metal halide lamps in several plants for an energy savings of 35 to 40 percent. In another case, the Hormel meat processing company upgraded its lighting system from incandescent lamps to a mixture of fluorescent and high-pressure sodium lamps for an annual electricity bill savings of $36,900, and the payback period for this project was calculated at slightly less than a year and a half. Lower-cost activities like decreasing excessive lighting levels and light fixture cleaning programs often have even shorter payback periods.

Another major area of industrial energy management practices focuses on improving the efficiency of plant steam generator and furnace use. A common furnace improvement technique is to increase the level of insulation used in the furnace to lower the amount of heat lost to the surrounding plant. For example, a Philips Industries plant achieved annual natural gas savings of $49,000 by replacing worn insulation on a high-temperature annealing furnace with flexible, crack-resistant insulation. In addition to providing higher levels of thermal insulation, the new material enabled the plant operators to turn the furnace completely off on weekends; previously it had been operated at low temperatures to prevent the old insulation from cracking. Substantial energy savings are achievable even with lower-temperature furnaces. An electric oven in a food processing plant was insulated with 2 inches of fiberglass material and operators reported a savings of 200,000 kilowatt-hours per year, or roughly $11,000.

Similar actions can be taken to improve the efficiency of a plant's boiler and steam distribution system. One of the first actions an energy manager can take in improving a plant's steam system is to locate and repair leaks in the steam traps which are used to remove air and condensed water from a plant steam distribution line. Nick Misra, corporate energy manager for the Dennison Manufacturing Company, says that a detailed inspection of the steam lines in his paper plant uncovered $63,000 in wasted steam produced by the 3 percent of steam traps in the plant that were faulty. He comments: "Any company that hasn't looked at its traps in the last five years is probably blowing about 20 percent of its energy costs down the drain because of malfunctioning traps."[8]

Under Misra's management, the Dennison plant has undertaken a comprehensive boiler and steam-system improvement program. The company reevaluated the thermal integrity of its steam distribution lines. This assessment resulted in the removal of 900 feet of uninsulated

[8] Tim Mead, "Paper Plant's Policy Shift Spurs Savings," *Energy User News*, Apr. 27, 1981, p. 8.

underground steam pipes, which were costing the company $14,000 each year. Automatic valve control devices and deaerator controls were installed to replace inefficient mechanical controls on the plant boilers. These controls are expected to save the company $9500 a year, according to Misra. The boiler improvement program also included the installation of two automatic devices: one to deliver the exact amount of heat required to hold the boiler fuel oil at the right viscosity and one device to recover heat contained in the blow-down liquid expelled from the steam lines to minimize the buildup of dissolved solids in the steam system. These devices saved the plant $6000 and $9000, respectively, in the years in which they were installed.

A final addition to the boiler system was the installation of oxygen analyzers and air-to-fuel ratio controls to insure the proper mix of oxygen and fuel in the boiler combustion chamber. Too much air in the system raises boiler energy demands because the excess air must also be heated to combustion temperatures, and too little air results in the waste of unburned fuel up the exhaust stack. The combustion control equipment increased the efficiency of the plant boilers by 3 percent and saves an estimated $126,000 in fuel costs each year. Combustion control equipment can also be used effectively with industrial furnaces and ovens.

SUBSTITUTION: TECHNOLOGIES THAT REPLACE FUEL USE

Because of the relatively high cost of combustion control systems (around $76,000 for the four controls installed in the Dennison plant between 1976 and 1978), some energy managers would classify them as equipment efficiency improvements that substitute capital for fuel costs rather than "housekeeping" investments. The primary goal of such substitution technologies is to tune energy equipment so that it converts raw energy into useful forms as efficiently as possible and then uses the converted energy as completely as possible.

In addition to combustion control devices such as the ones mentioned above, energy managers can often improve equipment efficiencies by installing equipment designed to remove useful heat from a boiler or furnace exhaust stack and put it to work elsewhere in the plant. When a boiler is fitted with an economizer—a waste-heat recovery device designed to remove heat from boiler exhaust gases in order to preheat the incoming feed water—efficiency improvements ranging from 2.5 percent to more than 10 percent are possible, depending on how hot the exhaust gas entering and leaving the device is (the larger the temperature difference between the entering and exiting air, the greater the amount of heat transferred to the incoming boiler water).

The Arcata Corporation installed a boiler economizer to recycle exhaust heat in their Waterville, Maine, container plant and realized $72,000 worth of annual fuel savings. The payback period for this type of investment is often quite low. When the Hammermill Paper Company installed flue gas economizers on two small package boilers in their Oswego, New York, plant, they saved $150,000 per year on their investment of $182,000.

Similar savings are possible using the waste heat in furnace and oven exhaust gases. For example, the Hormel Company has installed heat exchangers in their Ottumwa, Iowa, meat processing plant to capture vapors from cookers in order to provide hot water for plant cleanup. An additional benefit of the system is that it eliminates unpleasant odors from the cookers. The savings generated from this project are about $70,000 per year on an investment of only $60,000.

Industries which operate furnaces at very high temperatures are generally able to achieve higher savings than those which operate at lower temperatures because their exhaust gases have a higher thermal-energy content. For example, an Allegheny Ludlum Industries steel plant installed waste-heat recovery devices, or recuperators, on several old and inefficient steel furnaces and used the reclaimed heat to preheat cold steel strip entering the furnace. Additional recovered heat is used to heat the plant in the winter. Fuel inputs to the furnaces were decreased by 40 percent, with a corresponding increase in production of 50 percent. The cost savings generated by this project totaled approximately $456,000 per year. While this might be an extreme example, savings of 15 to 20 percent are easily obtainable even on newer metal-industry furnaces, and a large percentage of new furnaces now come already equipped with exhaust-heat recovery equipment.

PROCESS CHANGE

Using recycled furnace exhaust heat to provide plant space heating is also an example of a way that energy managers can achieve considerable energy efficiency improvements altering the flow or timing of industrial processes to make better use of the quality provided by each Btu of energy consumed. The quality of energy varies with its temperature because energy at higher temperatures can provide more useful work than energy at low temperatures. For example, energy at 1600°F has the capability of melting steel as well as making steam, while energy at 300°F could only manage the latter task.

Industrial energy productivity can generally be improved when managers match the quality of heat available in their plants to the chore at hand. For instance, energy managers might redesign the flow of products in their plants to recycle high-quality steam (at high temperatures and pressures) through several processes while it gradually lost quality as work was performed at each step. Experts believe that it is likely that the use of low-temperature steam boilers could be virtually eliminated in many plants by this method of cascading higher-quality steam through several plant processes until its energy content is exhausted.

Another form of process change involves the use of equipment that is totally different from that used in a plant's current production techniques. As energy prices began to skyrocket in the 1970s, newer, more energy-efficient industrial processes began to appear in almost all industries. One of the most common changes that has been initiated in many industries is the conversion from batch to continuous processing. Wood pulping, metal casting, and fabric dying and finishing processes have all been modified to allow continuous processing. The advantage of continuous processing is that it eliminates several heating-cooling or wetting-drying sequences in the processing of the end product. This lowers the final energy requirements to process a unit of the product. Continuous process savings can be substantial; for example, batch pulping requires roughly two times the energy input of continuous pulping. In addition to saving energy, continuous processes often allow for raw materials savings owing to less handling and fewer opportunities for product degradation.

In the chemical products industry, new methods of producing essential chemical reactions have been developed that can lead to substantial energy service savings. For example, PPG Industries has developed a proprietary method for making chlorine and caustic soda that uses 25 percent less energy than the present manufacturing technique. A conversion of their Lake Charles, Louisiana, plant to the new process is expected to achieve an electricity savings worth approximately $54 million per year. Similar savings levels are being achieved in the production of other chemicals with the use of more efficient production processes.[9]

Dramatic energy savings have also been realized in the cement industry through the implementation of a European manufacturing process. While conventional cement making involves the mixing and transporting of cement materials with water, the European process mixes and preheats cement materials without the use of water. This eliminates the necessity for the vast quantities of energy that are used in the United States to dry cement and results in energy requirements half of those demanded by the wet process.

A complete list of all the process changes implemented in this country

---

[9] "PPG Allots $100M to Cut Energy Use," *Energy User News*, March 23, 1981, p. 6.

since 1970 would take several pages, but these examples should indicate the abundance of energy productivity improvement opportunities available to innovative industrial managers willing to change their methods of doing business as usual.

COGENERATION

Another way of improving energy efficiency is cogeneration—the simultaneous production of useful process heat and electricity. Interest in cogeneration has been revived by the recent price increases in steam and industrial electricity.

The reason behind this interest is the much higher efficiency of cogeneration compared to the separate generation of thermal energy (usually steam) and electricity. While conventional power plants operate with efficiencies of around 33 percent for generating electricity, and new industrial boilers operate in the 80 to 85 percent efficiency range, industrial cogeneration systems produce both at total system efficiencies ranging from 60 to 80 percent—much higher than the average efficiency if the products are generated separately.

While approximately 50 percent of U.S. electric needs in 1900 were supplied by industrially based generators, this fraction has fallen to only about 8 percent in the early 1980s. Only part of this industrial-electricity generation was produced by cogenerators, while the rest used self-generation or on-site electricity-only generation plants. Estimates of cogeneration capacity installed in the United States in the early 1980s range from 9100 to 14,858 megawatts.[10] Assuming that the lower estimate is accurate and that cogeneration plants operate an average of 57 hours out of every 100, industrial cogeneration might have provided 45 billion kilowatt-hours of power in 1980, or roughly 5 percent of the total amount of electricity used by industry in that year.

Most analyses conclude that cogeneration's contribution will begin to grow. In the projections illustrated in Figure 5.1, industrial cogeneration could be increased from 5 percent of industrial-electricity use to about 11 percent by 1990 and 12 percent by 2000. These projections, which assume that barriers to the implementation of economic cogeneration projects are overcome, attempt to measure the number of economically attractive cogeneration projects. This estimate may actually be somewhat conservative, since our analysis only considered cogeneration in new or replacement industrial facilities. If replacements of existing boilers were

[10] Office of Technology Assessment, *Industrial and Commercial Cogeneration*, Washington, D.C., February 1983.

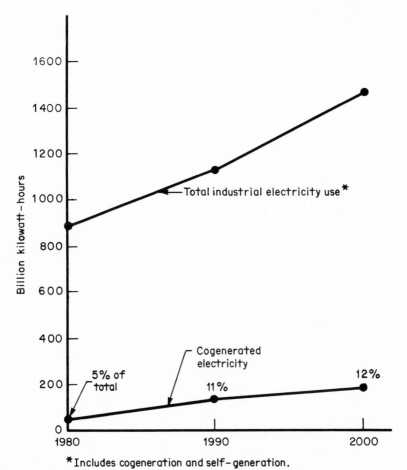

*Includes cogeneration and self-generation.

Figure 5.1 Cogenerated and total electricity use in industry. (*Source*: Applied Energy Services, Inc., Least-Cost projections.)

also included, the cogeneration potential would probably be more than double this estimate.

As shown in Table 5.4, the most competitive cogeneration system in the near future may use natural gas—most likely, a gas-fired turbine that generates electricity and then uses the exhaust from the turbine to make steam. Both the electricity and the steam can be used by the company, with any excess sold to the electricity grid. The economic attractiveness of the gas-turbine system depends heavily on the spread between electric and gas prices, however. In these projections, the differential between electricity and gas prices is large enough to make gas-turbine cogeneration competitive in many systems. Gas-fired cogeneration accounts for

TABLE 5.4 Least-Cost Cogeneration Capacity Additions (Megawatts)*

Fuel	1980–1990	1980–2000
Natural gas	11,200	15,300
Petroleum coke	900	1,300
Coal	2,800	6,700
Other	200	4,200
Total	15,100	27,500

* Additions to 1980 cogeneration capacity, estimated at 9100 megawatts; average capacity utilization is assumed to be 67 percent.
SOURCE: Applied Energy Services, Inc., Least-Cost projections.

almost half of combined steam and electricity production potential from all cogenerators indicated in our projections. However, should gas prices go higher than we estimated, cogeneration can be fueled by coal, and many other cogeneration equipment configurations might become attractive (such as coal, petroleum coke, or perhaps even water-slurry mixtures burned in large slow-speed diesel engines).

To give a comparison of the scale of conventional central electricity generation and cogeneration plants, it might take twenty-eight 1100-megawatt coal or nuclear base-load utility plants to generate the same electrical output as the 1500 industrial cogenerators (averaging about 20 megawatts each) indicated in the Least-Cost projections. The smaller cogeneration plants do have some planning advantages over larger plants. The lead time needed to plan and construct them is much shorter than for larger coal and nuclear plants. Their smaller capacity enables the electric system to adjust generation capacity to demand needs in smaller increments, eliminating the early years of overcapacity that often result when a huge central power generator is brought on line.

Even though cogeneration appears attractive because of its higher efficiencies and more manageable project sizes, we discovered in our research that some significant barriers must be overcome if its potential is to be realized. For example, under present law electric utilities are discouraged from investing in cogeneration as part of their business. The main reason for this is that electric plants are regulated; cogeneration plants are not. In order to qualify for the special legal standing of cogeneration and, therefore, of unregulated electricity, a plant can have no more than 50 percent utility ownership. Thus there is little incentive for utilities to enter the cogeneration business.

Secondly, businesses with potential cogeneration sites generally place a higher priority on production plant or revenue-increasing investments than they do on investments to reduce the cost of their utilities— electricity and steam. As a result, cogeneration projects are often put at the bottom of the capital-expenditure list. There are some significant exceptions to this rule (such as Dow Chemical and some paper companies), but the majority of companies we have talked with are deferring cogeneration projects in favor of more revenue projects. An executive of a large oil company told us there was no way a cogeneration project could ever be funded when there was so much capital needed for oil and gas exploration. Chemical, steel, aluminum, and other energy-intensive industry executives have made comparable comments.

In spite of these barriers, some companies are moving ahead with cogeneration projects on their own. For instance, International Paper Company has installed a new wood-waste plant at Gurdon, Arkansas, that provides 93 percent of their energy requirements. The cost of the plant was $30 million, and even though savings are not disclosed, the management is very pleased with the results. Also Blandin Paper Company in Grand Rapids, Minnesota, has built a $35 million cogeneration system using wood refuse—a by-product of its manufacturing process—and low-sulfur coal as fuel. Presently 55 percent of their energy output comes from wood refuse and they expect that to jump to 75 percent in the near future.[11]

As a part of this scenario a new industry is emerging—third-party businesses that own and operate unregulated power and steam generators. These entities are building cogeneration plants, then selling the electricity to some power-short utilities and the steam to a single industrial customer. Only a few of these plants have been completed, but the number of projects pending is growing. Applied Energy, Inc., until recently a subsidiary of San Diego Gas and Electric, is furthest along with some sixty megawatts in service. Several other companies, including our own, have also announced projects that total from 3000 to 5000 megawatts. In one such 150-megawatt AES project at a large Houston refinery, the electric utility will buy the electric output of the cogeneration facility at a price equal to their natural gas savings and the industrial customer has agreed to buy their steam at a price that equals roughly 40 percent of its projected steam costs. Even with these low prices, the new cogeneration utility is expected to obtain a 15 to 18 percent after-tax rate of return for its equity investors.

---

[11] Several other examples are worth noting. Diamond Shamrock, Inc. erected a 4.5-megawatt facility in Stockton, California, that became operational in 1980. It is fueled by walnut shells and saves the company approximately $1 million per year. Paul Masson Vineyards installed a 290-kilowatt natural gas–fired system in Saratoga, California, that provides all plant hot water and one-third of plant electricity demands. The savings are roughly $65,000 per year on an investment of $175,000. Hammermill Paper Company installed a steam turbine in Hamilton, Ohio, to drive a paper-machine line shaft, replacing an electric motor. The steam comes from a 150-psig source that was previously throttled back to 40 psig for plant needs. Annual savings of $60,000 per year are approximately equal to the investment.

USE OF LOW-COST OR NO-COST FUELS

The Blandin Paper Company cogeneration plant also illustrates another trend in the movement to decrease the cost of energy services. Many plant managers are finding that they can make use of low-cost fuels like wood, coal, or petroleum coke in their processes, especially in their boilers. New technologies like fluidized bed combustion allow industrial users to consume coal or other dirty fuels without endangering the environment or spending as much on conventional pollution control systems.

Other companies are finding ways to make use of previously wasted materials. For example, the Armco steel plant in Ashland, Kentucky, replaced an oil and natural gas–fired boiler with one that can burn the plant's byproduct blast furnace gas, which was previously flared and wasted. This move has saved the company $250,000 per month.

In addition to steel and paper industry plants, food product manufacturers are finding the use of waste materials for generating energy services very attractive. A Gold Kist plant in Georgia installed a boiler capable of generating 70 percent of the plant's steam demands with peanut shells from a nearby plant for 10 months of the year. The boiler will fire waste wood from sawmills for the remaining months of the year. The cost of the boiler and fuel handling system is nearly $4 million, and the project is expected to pay for itself in 3 years.

INDUSTRIAL LEAST-COST PROJECTIONS

What level of energy consumption is likely in the future if these trends to improve the productivity of industrial energy use continue? The analysis we conducted, summarized in Figure 5.2, indicates that industrial energy use could be 35 percent lower than it otherwise would have been in the year 2000 from conservation and efficiency improvements. These savings come from a combination of factors: changes in output mix to less energy-intensive products within each industry and the continued implementation of energy management, process changes, and process efficiency improvements. Process change accounts for the largest part of these potential improvements, while the rest of the savings are divided about equally between change in output mix, energy management, and process efficiency improvements (like the use of waste-heat recovery devices).

The total energy use resulting from our Least-Cost projections is shown in Figure 5.3. While industrial production is projected to increase at 3.3 percent per year during this period, energy use might increase at only slightly over 1 percent per year. Because of its relatively high price,

practically no new oil-fired facilities are likely to be built over the next
two decades. Consequently, oil use should decline to 80 percent of 1980
consumption levels by the year 2000. With the increased use of advanced
coal-capable technologies, we think coal use will increase at over 5

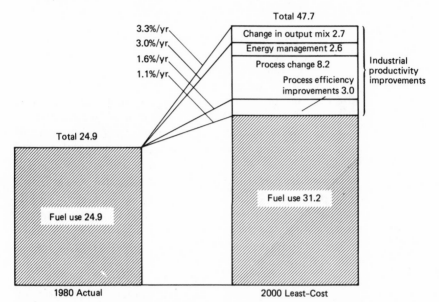

Figure 5.2   Projected industrial energy consumption (quads per year).

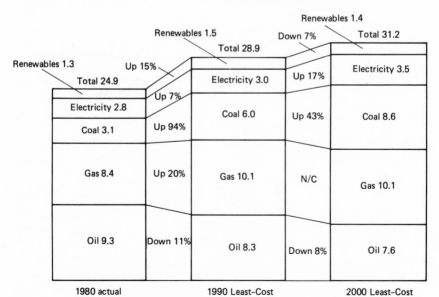

Figure 5.3   Least-Cost industrial sector fuel use (quads per year).

percent per year, resulting in consumption by 2000 nearly three times its 1980 level.[12]

IS THIS FUTURE REASONABLE?

Are these projections realistic? If our price and economic assumptions prove accurate, and managers in the future continue to make use of the many opportunities available to them to improve the productivity of energy use in their plants, our projected industrial-energy future will easily be realized. Although no one knows exactly how much further industrial-energy productivity can be improved in practice, the American Physical Society indicated in their 1974 energy conservation study that it is possible to achieve an 80 percent reduction in the 1972 energy-to-industrial-production ratio.[13] This would suggest that industry has achieved only one-quarter of the potential improvement and will still have another 40 percent to go in 2000 if our projections are realized. The argument that not much more can be done to save energy in the industrial sector beyond the 15 or 20 percent improvement already achieved is a myth. In practice, energy productivity is providing a powerful means to improve product costs, rather than an option of last resort. Industrial-energy systems, along with those in buildings and transportation, are proving to be adaptable enough to provide an abundance of energy service options at prices equal to or less than present systems.

---

[12] At present, our models limit the use of renewable fuels like wood and food industry by-products to the paper products industry, so our estimate of industry-renewable fuel use is probably conservative. In addition, these projections were based on possible modifications to new and replacement equipment only; if we had included the improvements that can be made to existing equipment, the fuel use projections would likely have been even lower.

[13] American Physical Society Summer Study of Technical Aspects of Efficient Energy Utilization, 1974. Available in W. H. Carnahan et al., *Efficient Use of Energy, A Physics Perspective*, from the National Technical Information Service (NTIS) (PB-242-773), or in *Efficient Energy Use*, vol. 25 of the American Institute of Physics Conference Proceedings.

6

Alternatives to Traditional Fuels

The potential quantities of alternative fuels are ample but for the most part, costs are not competitive.

Discovering the benefits of energy productivity has been one of the most important developments in recent years. As we have seen, the major cost-saving energy alternatives are associated with the use of energy, but there are some exciting alternatives to traditional fuels emerging as well.

Until the late 1960s, alternatives to traditional fuels were of little interest to anyone but a few scientists. In retrospect that is understandable, since traditional sources were abundant; their costs were declining, while the quality and quantity of production were on the rise. Therefore, most of the important developments in alternative fuels occurred only in the last decade, when the cost of traditional fuels began to rise.

TRADITIONAL FUELS

Before reviewing alternative fuels it is important to briefly examine the status of the traditional fuels that America has relied on for the last 30 years—oil, natural gas, coal, and electricity—the fuels that, when they were cheap and abundant, created the most powerful economy the world has ever known.

Oil

In 1982, the U.S. economy consumed a little over 30 quadrillion Btu of refined petroleum products, 43 percent of all the energy consumed that year. That was down 20 percent from the all-time peak experienced in 1978 and was almost equal to 1970 oil use. Other nations supplied 29 percent of the U.S. 1982 oil demand, sharply down from the peak of 49 percent in 1978 and nearly back to 1970 import levels of 23 percent.[1]

What are the expectations for the future? Although there is disagreement about how much oil the United States will want to use, there is some consensus about how much can be produced domestically. At best, U.S. oil production, which now represents 20 percent of world output, will remain constant for a few more years and then begin a long and continuous decline. By the year 2000, the quantity of petroleum produced in the United States, including Alaska, might be from 5 to 20 percent below the current supply, depending on its price over the next few decades.[2]

In contrast, production in the oil-exporting countries is running well below current available capacity. The Organization of Petroleum Exporting Countries (OPEC) produced only about 35 percent of the world supply in 1982, some 30 to 40 percent below their present capacity. Oil use per unit of world output declined 24 percent between 1978 and 1981, as OPEC priced their oil above the energy market. The current price of $29 per barrel, down from the 1980 peak of $40, brings it more in line with other fuels but is still far from restoring OPEC to desired levels of production. Even at $29 per barrel oil is priced out of most stationary markets—buildings, industry, and utilities—and will remain competitive only in the transportation market. Without debating OPEC's objective, it is clear that their actions created a large cushion in the world oil market for years to come.

But even though an oil crisis is much less likely, and we expect that world oil prices may continue declining in real terms for the next several years, the gradual long-term decline of oil availability suggests that a resumption in real oil price increases is inevitable.

Natural Gas

The supply of natural gas is perhaps the least certain of all U.S. energy sources. While the consumption of other fuels has risen, the United States is using less gas now than it did in 1970 (about 18 quads in 1982

---

[1] Data from Energy Information Administration, *Monthly Energy Review*, Aug. 1983.

[2] See, for example, U.S. Department of Energy, *Energy Projections to the Year 2000*, July 1982, or Energy Information Administration, *1981 Annual Report to Congress*, vol. 2, February 1982.

versus 22 quads in 1970). In 1982, natural gas accounted for only 26 percent of total energy consumed, down from 33 percent in 1970.

The problem is that gas prices have been held artificially low by federal price regulation since the early 1950s. Therefore, as gas prices are deregulated, it is difficult to predict how much gas can be produced at higher price levels. And since gas has strong competition in almost all of its uses, especially homes and factories, prices strongly affect demand.

Even the most optimistic projections of future gas supply assume a decline in the traditional sources of cheap gas. The Gas Research Institute's 1982 baseline projections showed 15.8 quads of conventional gas in the year 2000, a decline of some 25 percent below 1980.[3] The American Gas Association's projections of conventional gas are even lower, a total of some 12 to 14 quads.[4] Most of the new gas supply sources being counted upon by the gas industry are unconventional, with the exception of an increase in the supply from Mexico. While the potential quantities of those new sources of gas are important and will be described later, proven reserves of conventional natural gas have declined steadily from about 14 years supply in 1970 to 10 years in 1980.[5]

Coal

After years in the doldrums, primarily because of competition from artificially cheap and more convenient natural gas, coal is regaining some popularity. During the 1970s, the coal mined and consumed increased 35 percent, which represents an annual rate of a little more than 3 percent per year. From 1975 to 1980, coal consumption increased at an even faster annual rate of 4.4 percent per year while the total domestic production of other fuels decreased 0.3 percent per year. Almost all of the increase in coal consumption over this decade was for coal used in electric utilities.

Even with these increases in coal sales, the potential supply of coal in the United States still far exceeds demand. Excess coal-mining capacity and the ease of opening new mines makes coal our most abundant energy resource. Coal is so abundant that in those same 5 high-growth years (1975–1980) the price of coal declined (in constant dollars), while

---

[3] "Baseline Projection of U.S. Energy Supply and Demand", Gas Research Institute, Dec. 23, 1981.

[4] George H. Lawrence, *Testimony before the Subcommittee on Fossil and Synthetic Fuels of the Committee on Energy and Commerce*, U.S. House of Representatives, Aug. 6, 1982, p. 26.

[5] As measured by the reserve-to-production ratio. A decline in this ratio indicates that reserve additions have totaled less than production. Data on gas reserves and production are from the Energy Information Administration, *1981 Annual Report to Congress*, vol. 2, February 1982.

the composite price of all other fuels rose 71 percent. This supply surplus is likely to continue for some time, for the United States has vast resources of coal to tap. Mapped and explored U.S. coal resources could last for about 2300 years at current rates of consumption. Moreover, recoverable proved reserves alone would account for something over 300 years supply at the current rate of consumption.[6]

Within the coal industry, there is ample competition to keep prices down, even when natural gas prices are totally deregulated. As of 1980, the ten largest coal-mining companies controlled only 33 percent of the market, as opposed to 54 percent for oil.[7] As a result, coal is likely to keep all other fuel prices in a rational balance. Oil has already been priced out of many of the markets that coal serves—including electricity made from coal-fired steam plants—and gas may suffer the same fate if its price rises too quickly. Furthermore, coal can be converted into oil or gas equivalents, which have the potential of placing a long-term ceiling on gas and oil prices (admittedly, a relatively high price ceiling). Therefore, for Americans, coal is an effective anchor on fuel prices. There is enough competition in coal mining that the major factors influencing coal prices will likely have more to do with the costs of transporting and handling coal rather than with coal mining. It will continue to be a cost-based fuel, since shortages are not likely to occur long enough for market prices to rise significantly above costs.

Electricity

Electricity is somewhat different from the other fuels discussed here in that it is generated by combustion turbines or steam from boilers fueled primarily by oil, gas, coal, nuclear, or hydro power. In the conversion of fuels to electricity, a good deal of energy is lost. It takes about 3 Btu of fuel to make 1 Btu of electricity. Because electricity is clean and convenient and can generally be used in higher-efficiency end-use equipment, consumers are willing to pay for its extra cost.

Sales of electricity totaled about 2 trillion kilowatt-hours per year in 1982.[8] That is the equivalent of keeping eleven 100-watt light bulbs burning the year around for each person in the United States.

As shown in Chapter 1, until the late 1960s the cost of building new electric power plants declined in constant dollars, and as a result the price of electricity declined as well. Since that time, however, there has been a dramatic turnaround in construction costs, and the cost of

[6] Energy Information Administration, *1980 Annual Report to Congress*, vol. 2, p. 41, 103.
[7] Energy Information Administration, *1980 Annual Report to Congress*, vol. 2, p. 129.
[8] Energy Information Administration, *Monthly Energy Review*.

electricity from new power plants has increased by roughly 150 percent.[9] As these higher power costs are passed on to consumers, the demand for power has slowed to a 0 to 2 percent per year annual growth rate, compared to the historic (pre-1973) growth rate of 7 percent per year. Since no one planned for such low growth rates, many new plants were started in the 1960s to meet increases in demand that never appeared. As these new plants were completed, the reserve margin (the percent of unused capacity) grew from 19 percent in 1970 to about 40 percent in 1982.[10]

In 1982, 52 percent of the electricity used in the United States was generated by coal, as compared to 46 percent in 1970. Of the remainder, 15 percent was generated by natural gas, 11 percent in hydroelectric plants, 12 percent by nuclear, 9 percent by oil, and 1 percent from geothermal, wood, and water products.[11] But the real question is how this mix of sources might evolve in the future.

Our cheapest electric power option is electricity supplied from hydroelectric dams that provide the water flow to run turbine generators. Electricity of this kind has always been extremely cheap and is now roughly one-fifth to one-eighth the cost of conventional power plants. Most of the hydropower resources are concentrated in the Pacific northwest and the Tennessee Valley. The difficulty is that the nation has run out of large dam sites. What remains are sites for small hydro plants, capable of producing only 10 to 50 kilowatts, a small fraction of the unit size of the major plants now in use and, in total, a small fraction of the current production.

Many utilities began major nuclear and coal power plant construction projects in the 1970s to meet anticipated increases in demand. Yet as these plants proved to be more expensive than expected, electricity prices rose, and the demand didn't materialize. As a result, many of these new projects—especially partially completed nuclear power plants—are being cancelled or postponed. The financial realities of nuclear power have done what protesters never could—stopped the development of new plants. No applications to build a new nuclear

---

[9] Data from Charles Komanoff, *Power Plant Cost Escalation*, Komanoff Energy Associates, New York, 1981, p. 2. Nuclear capital costs escalated 142 percent from 1971 to 1978 (12.6 percent per year); coal plants (including scrubbers) escalated 68 percent (7.4 percent per year). Assuming costs continued to escalate from 1978 to 1982 linearly, nuclear costs are inferred to have increased 220 percent from 1971 to 1982, and coal 107 percent. Since new orders were split roughly equally between coal and nuclear plants in this period, capital costs for new plants escalated about 160 percent on average over this period.

[10] North American Electric Reliability Council, *Electric Power Supply and Demand*, 1982–1991, Princeton, August 1982, pp. 43, 44.

[11] Energy Information Administration, *Monthly Energy Review*.

power plant have been filed with the Nuclear Regulatory Commission since 1978, and numerous other projects have been dropped or deferred.

In 1980, 17 reactors were cancelled and 70 deferred. In 1981, 6 were cancelled and 44 deferred. And in 1982, 19 were cancelled and 42 deferred. Even the gigantic Tennessee Valley Authority, which is government-owned, has had to cancel 4 of its 14 plants under construction and defer 4 others.[12] The cost of nuclear power plants has increased nearly ten-fold since 1970, prompting one industry executive to say, "In my opinion, we really don't have a nuclear option."

For coal-fired power plants, there are enough sites, coal mines, engineering talent, and materials to meet almost any necessary schedule of construction. The issue of how many coal plants to build is therefore a function of demand, not supply. The question is whether the power generated from new plants can be sold in competition with other alternatives. And if the power is marketable, can the capital be raised by the utilities?

With current reserve margins almost double the desired levels, a 15 percent increase in electricity demand could theoretically be met without even finishing the plants under construction. If all plants that have been started were completed, then roughly a 30 percent increase in demand could be accommodated from the total system. Shortage, therefore, is not the issue for electricity. More at issue is the cost and marketability of electricity (in competition for energy services) in current markets.

RENEWABLE ENERGY SOURCES

Renewable energy sources—also loosely categorized as solar energy—are the alternative fuel sources that truly excite many Americans. Although solar energy is generally thought of simply as sunlight, its derivatives embrace a number of other resources that are continually renewed by the natural processes emanating from the sun. In a broad sense these renewables include hydroelectric dams, heat collected by windows and greenhouses, methane or liquid fuels from garbage or biomass products, electricity from windmills, thermal collectors that magnify the sun's radiation, sophisticated photovoltaic cells that create electricity, and wood.

Renewables are popular because they are clean, resilient to disturbances, and appropriate (well-matched to the energy services provided), and the fuel source (sunlight or its derivatives) is free or at least low in

[12] North American Electric Reliability Council, *Electric Power Supply and Demand, 1982–1991*, Princeton, August 1982, pp. 43, 44.

cost. Yet the cost of purchasing, installing, and maintaining the equipment needed to deliver renewable energy services can be significant, making the choice between a traditional or renewable system complex. Some renewable options—particularly passive solar designs and some solar collector applications—are cost-effective choices now, especially after a home owner has installed the types of energy efficiency improvements described in earlier chapters. If conventional fuel costs escalate more rapidly, other renewable options might be developed to provide substitutes for these fuels, and ultimately to moderate the cost of energy services to consumers.[13]

Photovoltaics

The photovoltaic effect of the sun has been of interest for more than 100 years, and research has shown that the possibilities are phenomenal. The technology was initially developed in the early years of the U.S. space program, as an energy source for spacecraft. Photovoltaic cells produce direct electric current from sunlight, with no chemical reaction and virtually no maintenance. In addition to their space applications, they are ideal for remote earthly applications such as highway signs, buoys, remote microwave relay stations, or desert military bases. Believing its potential to be large, oil companies have bought up major photovoltaic firms and are investing large sums in its development.

At present, the electricity generated from photovoltaics is more expensive than conventional sources of electricity by about a factor of 10. However, as occurred with transistors and microcircuits, the cost per watt of photovoltaic cells has fallen dramatically since their introduction. The cost of photovoltaic cells made from silicon wafers has dropped from $30 per peak watt in 1976 to between $7 and $10 per peak watt in 1979–1980. Second-generation processes that manufacture the cells more cheaply are under commercial development, and the U.S. Department of Energy has the goal of reducing the array price to a level comparable to the cost of electricity generated from a new central-station power plant by 1986. In addition, many small firms are now offering photovoltaic cogeneration systems, which concentrate sunlight on an array of photovoltaic cells and use the waste heat for domestic purposes. These systems may be cost-effective in some homes now, depending on the price of electricity and heating fuels at the proposed site.

[13] We are indebted to Amory Lovins and Hunter Lovins for their helpful suggestions on this discussion of renewable resources, including the information taken from Appendix 2 and 3 of their book, A. B. Lovins and L. H. Lovins, *Brittle Power*, Brick House Publishing, Andover, 1982.

Solar Collectors

Both passive (with no moving parts) and active solar collectors have become both practical and popular systems to reduce or eliminate the need for conventional heating systems in many homes. Passive solar designs—where solar energy is collected and stored by the building shell itself—have proved to be extremely cost-effective, especially when combined with building-shell efficiency improvements. New materials for window glazing, which insulate and reflect heat back into the building, are now in commercial production.

Active solar heating systems, which many still think of as ugly steel-and-plastic contraptions on rooftops, have become much more attractive, sophisticated, and technologically advanced. New plastic films have been developed for active collectors that are cheaper and more durable than conventional materials. More than 350 U.S. companies now make active solar collectors of both flat-plate and concentrating collector designs. The efficiency of flat-plate collectors can now be greatly increased by the use of "selective" surface coatings that absorb solar heat even on cloudy days.

With these advances in designs, and with continued advances in materials and collector systems, active solar collectors could be a competitive supplement or alternative to conventional heating systems in many installations. However, to be Least Cost, active systems must be properly designed and integrated into the building to minimize the need for extra expenditures on backup conventional heating systems or expensive heat storage equipment.

Wind Power

New and efficient machines have been developed to generate electricity from wind. Wind power was used extensively as a source of mechanical energy in the 1800s and for power in the 1900s, but its use evaporated with the introduction of steam-electric generators. The fuel prices of the 1970s stirred up renewed interest in this technology. As a result of extensive government-sponsored research and the Wind Energy Systems Act of 1980, $900 million has been spent by the government to develop cost-effective wind systems and to increase its use in large-scale power generation.

The new wind machines do not look like their earlier cousins and operate much differently. A good modern design extracts about 30 percent of the power in the wind, with better power returns under normal operating conditions per square foot of installation space than solar cells. A simple machine in a windy site—parts of New England, the

Great Plains, or on the Pacific coast or islands—can produce electricity at about 5 to 15 cents per kilowatt-hour (depending primarily on the wind conditions at the site). This makes wind power a possible competitor with conventional electric systems for new load. The amount of wind generation actually installed will depend on the need for new power sources in regions with wind potential, and the cost of alternative sources (including conservation).

Biomass

Biomass and wastes—consisting of wood and wood wastes, municipal and industrial solid wastes, sewage, agricultural wastes, alcohol fuels, and landfill gases—currently supply a little over 3 percent of U.S. energy needs; most of it is wood and wood wastes. In contrast, the 1980 World Energy Conference estimated that the ultimate annual world potential biomass production is about 9.5 times the current annual world energy production. The reason for this seemingly contradictory condition is cost: the expense of recovering all the potential energy from biomass is generally prohibitive.

Nevertheless, there is considerable work in progress that could change those costs, especially after the year 2000. Numerous small-scale test plots are being evaluated in different regions of the country, selecting trees that have a rotation of less than 20 years and hence, an energy yield that is three to four times greater per acre than current forests. Several nonwoody herbaceous plants, notably the Jerusalem artichoke for alcohol production and the jojoba shrub, from which oil is extracted, show considerable long-term promise as high-yield biomass sources. Various aquatic species, particularly giant California brown kelp, which can be converted to methane (natural gas), are being tested as commercial biomass sources. These sources all have certain advantages: there are no competitive uses for the species, growth rates are rapid, yields are often high, and, for marine biomass, land availability is not a constraint.[14]

Biomass products can be used to produce gas, liquid fuels for vehicles, or electricity. The most prevalent application is the use of wood wastes for generating or cogenerating steam and electricity in the pulp and paper industry. Also, in 1982 there were 3 new wood-fueled electric power plants under construction, as well as 26 solid-waste energy recovery plants in operation and 24 more under construction. Extensive research is also being conducted in anaerobic digestion of wastes for the

[14] Donald L. Klass, "Energy From Biomass and Wastes: 1981 Overview," *Energy Topics*, Mar. 15, 1982, p. 3.

production of methane, thermochemical gasification of biomass and wastes, and hydrogen production from biomass.

Similarly, methods of producing ethanol and methanol and nonalcohol liquids from biomass are constantly being improved. Ethanol is produced primarily from crops, particularly corn. Methanol is also referred to as wood alcohol, and wood is probably one of the most undermanaged and underutilized resources in the United States. The United States has twice as much unused woodlands as fertile farmlands. With proper planning and maintenance, that resource could provide readily usable fuel forever. If oil and gas prices were to rise as they have, biomass-based liquids and gases might become cost-effective sources.

NONRENEWABLE ALTERNATIVES

In addition to renewable energy sources, there are a number of alternative fuel sources that are unconventional but depletable, or nonrenewable, in nature. These include unconventional gas, synthetic fuels, shale oil, petroleum coke, fuel cells, and new nuclear options (breeders or fusion). All of these supply technologies might be characterized as centralized, which means they are usually generated on a large scale at a central plant and can be distributed to customers through existing transmission and distribution systems (pipelines, electrical grids, and retail outlets).

Synthetic Fuels

Perhaps the most prominent of these alternatives are synthetic fuels— coal that is converted into gas or oil products using a number of chemical processes. So far, however, the cost of synthetic fuels is prohibitive—2 to 4 times the 1982 price of marketed oil and gas. Although these high costs make synfuels uncompetitive at today's market prices, technological improvements and rising conventional fuel prices might make synfuels competitive in the longer term. The first commercial-size U.S. coal gasification project, the Great Plains facility in North Dakota, is now under construction because of a loan guarantee from the Department of Energy and might well provide some very useful data about the costs of future plants. However, the facility is under the constant threat of cancellation because of the current gas glut.

Even though coal gasification can be an end in itself, another useful fuel emerges with the addition of one more catalytic process to the gasification, namely, methanol. Methanol is a very simple carbon-hydrogen molecule known as alcohol. It is the second simplest of the alcohol compounds and in many respects is similar to ethanol or grain alcohol. The principal difference is that ethanol can be used as both a fuel or a beverage, while methanol is used exclusively as a fuel.

As partially discussed in Chapter 3, methanol stands out as a potentially attractive fuel candidate from coal because of its low cost. It is estimated to run about half of the estimated cost of petroleum products from other coal liquefaction technologies and is marginally competitive with conventional gasoline. It is also suitable for distribution and use as a liquid, like gasoline. Methanol has other advantages as an internal-combustion-engine fuel. It is a clean-burning, high-performance, high-octane substitute for gasoline. Although its energy density is less than gasoline, its inherent fuel economy nearly offsets its volume disadvantage. Fully 20 to 40 percent gains in fuel economy can be achieved by using methanol in high-compression engines. Gasoline engines are being converted to methanol in several corporate automobile fleets in California with results that, so far, are promising. For years methanol has been the preferred fuel in the Indianapolis 500, owing to its high-performance characteristics.

There is enough coal in the United States to power automobiles with methanol almost indefinitely, but achieving that level of production is no simple task. A facility to produce 50,000 barrels per day of methanol, or roughly enough fuel for 1.5 million people, would cost in the range of $2 billion and probably take 5 to 8 years start-up time. Interestingly enough, however, methanol can also be easily made from natural gas, a possible use of the remote natural gas in fields such as Prudhoe Bay, Alaska, and offshore of the United States, Indonesia, and the Middle East. In addition, any oil well usually contains natural gas that is often unusable and wasted by being burned off at the site.

What this means is that there is a potentially attractive use for a considerable amount of natural gas which has not made it to the marketplace, primarily because of the high cost of transporting it from the wellhead to the market. For some remote gas fields it is not technically or economically feasible to build a gas pipeline; but in all such cases, if the price is right, it is possible to convert methane to methanol and transport it by ship to transportation markets. In effect, it is not necessary to wait 10 years to get our first tankful of methanol from coal; it could be available in 3 to 5 years if natural gas was first converted to methanol. As more methanol is needed, other exotic gas resources could be tapped, or coal-based methanol plants could be built.

The economics of methanol at $1.60 per gallon of gasoline equivalent versus $1.20 for gasoline is relatively simple. The methanol costs of about 30 percent more than gasoline are offset by a 30 percent improvement in fuel economy. On a mass production basis, a methanol-powered car should not cost any more than a conventional gasoline-powered car. Thus, if all the distribution facilities were in place, the methanol would become a competitive alternative to gasoline. If gasoline prices rise beyond our expectations, it could some day be the preferred choice.

Shale Oil

Another contender as a petroleum substitute is shale oil. Shale oil (also called kerogen) is contained in rock formations in Colorado, Wyoming, and Utah, and the United States has plenty of it. Research for converting shale to a liquid fuel has been conducted for 50 years, but production has been limited to research facilities or pilot plants. The most attractive characteristic of upgraded shale oil is that it makes an excellent diesel and jet fuel, and those fuels are likely to be in increasing demand in future years.

Shale oil can be recovered through three basic processes: surface retort, in-situ, and modified in-situ. Surface-retorting methods are rather straightforward. The rock is mined using relatively traditional mining procedures (either open-pit or underground). The rock is crushed and heated, and the resulting vapor is collected, condensed, and treated. The result is a barrel of thick, viscous kerogen, which can be fed into a standard petroleum refinery as an additional feedstock to produce jet fuel, diesel fuel, and gasoline.

In situ processing is done without mining the shale. Shafts or tunnels are bored into the shale formation, and explosives used to break up the rock underground. Heat is applied, or the shattered rock is ignited to retort the shale. The modified in-situ process utilizes both above-ground and in-situ retort for a particular shale formation. This process allows more control over the in-situ retort, at the same time utilizing some shale that was removed from underground.

The estimated cost of producing gasoline from shale is $1.50 to $1.80 per gallon (in 1982 dollars), some 50 percent higher than current market prices for gasoline. As a result, most companies have postponed their plans for exploiting this resource for an indefinite period. Only Union Oil is continuing activity in developing a commercial process.

Unconventional Gas

Next in line are a number of unconventional gas sources that could supplement declining conventional natural gas supplies. These include gas from tight-sands deposits in the western states, Devonian shale in the east-central states, and deep, high-pressured gas mixed with water (geopressurized gas) in Louisiana. The future costs and availabilities of each of these sources are uncertain, but tight sands and Devonian shale now provide 4 percent and 0.5 percent, respectively, of the nation's gas supply. Many believe those quantities could double by 2000 if gas prices rise sufficiently. There are also estimates that geopressurized gas production could range from as little as 2 years of current consumption to as much as 100 years. Much is unknown about this resource, however, so that little extraction is expected before the year 2000.

To provide longer-term gas supply alternatives, some fascinating research is being done on gas located deep in the bowels of the Earth. As the president of the Gas Research Institute, Dr. Henry Linden, says:

> Still farther in the future lies the possibility of near endless supplies of methane from deep in the Earth. If Thomas Gold and others who are measuring the occurrence and magnitude of methane seeps from tectonic faults are correct, there may be enormous amounts of methane still available to us as a legacy from the Earth's creation. The theory of abiogenic methane, simply described, entails the slow outgassing of deep gas that was incorporated into the Earth at its formation. This aboriginal gas is distinct from the methane that was formed by the bacterial decay of plants and small organisms. According to this new theory, earthquakes, volcanic eruptions, and a variety of other phenomena point to a spasmodic release of deep gases, of which methane appears to be the major component. Very few geologists now deny the existence of abiogenic methane; the issues are its quantity and recovery. The implications of this hypothesis, however, are so enormous for the long-term supply of gas that the theory needs to be validated or disproved.[15]

Petroleum Coke

Unlike the other nonrenewable alternatives, petroleum coke is basically a waste byproduct of crude oil refining. It is produced in the refining process after all the gasoline, kerosene, gas oils, lubricating oils, and other light products have been distilled from the crude oil. Pumps force the heavy residual oil that remains through tubes of an oven, where it is heated to a temperature of 1800–2000°F. Vapors are driven off and sent to a fractionator for separation into more valuable gas, gasoline, and gas-oil products. The solid coke remaining is removed by fuser rods, quenched by water, screened for size, and stored. It is a hard grayish-colored substance of almost pure carbon, which produces intense, smokeless heat when burned and contains little or no ash.

---

[15] Henry R. Linden, "Gas Technology Needs: Near-, Mid-, and Long-Term," paper presented at the Annual Meeting of the Institute of Gas Technology, Nov. 12, 1981, p. 5.

Nationally, about 25 percent of all coke produced is purified and fabricated into anodes and electrodes for use in electric furnaces and other electrochemical processes. It is also used as a fuel, but its low volatility has made it difficult to combust without blending with other fuels, and its generally high sulfur content (the refining process tends to concentrate much of the sulfur from the crude oil in this terminal product) has made it environmentally troublesome as a major fuel source.

Current U.S. production of petroleum coke is not large by national energy accounts—about 350,000 barrels per day, which is equivalent to 2 percent of oil consumption. Sixteen percent of this production is exported to Europe and Japan, while much of the remainder is used in the United States for new fuel purposes. A number of refiners in the south and southwest are currently adding coking capacity so that in the next several years there should be a 20 to 50 percent increase in capacity. Yet because of its lowly status as a "junk" fuel, it is very cheap. Accordingly, the long-term outlook for coke as a supplemental fuel source is favorable, both as to availability and price.

Nuclear Fusion

Although, as we have shown, the need for new electric generation sources is expected to be limited, the search continues for cheaper methods to generate electricity. Nuclear fusion is the most prominent of the exotic technologies being explored as potential long-term (post-2000) sources.

Fusion power is the same process that creates heat and light from the sun, and that powers most stars. The fuel is deuterium, or heavy hydrogen, contained in ordinary water. To liberate this energy, however, the sun's temperature—about 100 million°C—must be reproduced.

Many governments (primarily the United States and USSR) are spending large sums on fusion research. In the United States alone, over $20 billion will be spent on fusion research if current research budgets are followed, to overcome the barriers to commercialization of fusion power. The most important is the need for stable yet extremely high plasma temperatures. Depending on the results of these research and development programs, a commercial-scale fusion reactor might be operable by 2020 or 2030, if needed.

Breeder Reactors

Another nuclear technology is the breeder. While uranium is the fuel used in conventional nuclear plants, breeder reactors operate on plutonium and convert uranium to plutonium faster than they use plutonium as fuel. They thereby create more fuel than they use while producing energy. The surplus plutonium can be reprocessed for use in other breeders or even in conventional nuclear power plants. Advocates believe that the breeder could extend uranium resources 50 times over current estimates.

A number of issues cloud the breeder's future, however. The fuel cycle is complex and capital-intensive. Environmental questions are still a major consideration, but concerns of nuclear weapons proliferation, the danger in transporting the fuels, and cost overruns in construction and technical-engineering problems are delaying its commercial development in this country. Most importantly, lower growth in electricity demand has slowed the depletion of uranium to the point where the breeder may not be needed as a supplemental fuel source for another 100 years or more.

Fuel Cells

A technology along very different lines is the fuel cell. The fuel cell is an ingenious device that converts fuel to hydrogen, which is then chemically reacted with oxygen to produce electricity and usable waste heat. Developed during the space program in the mid-1960s, fuel cells were expected to provide industry with an alternative to conventional electric systems. They have the advantage of generating electricity with little chemical pollution, minimum noise, and no moving mechanical parts. Also, because of their relatively small size their waste heat can be used on-site.

An advantage of the cell is that unlike some standard generators that lose efficiency if not performing at full capacity, fuel cells retain their efficiency even at partial loads. Small-to-intermediate-sized fuel cells have been tested in some commercial buildings. In smaller sizes, they provide the electricity and space conditioning for entire buildings and are sometimes called "total-energy systems." In larger sizes, they provide highly efficient electric subgenerating plants. One 8-megawatt operation has recently begun in New York City with the cooperation of Consolidated Edison, and gas companies are evaluating some fifty 40-kilowatt total-energy systems.

Today's fuel cells are designed to run on hydrogen generated from natural gas, and an assured supply of reasonably priced natural gas is therefore necessary for their commercial operation. But work is being done to develop cells that run on other (perhaps cheaper) fuels like coal, naphtha, alcohol, propane, petroleum, or methanol.

CONCLUSION

Just as earlier chapters have demonstrated an abundance of cost-effective efficiency improvements, this chapter suggests a similar abundance of alternative fuels. Because energy efficiency improvements are by and large the Least-Cost choice in most energy service categories, they will likely be deployed first.[16] As these investments in efficiency improvements bear fruit, the future demand for both conventional and alternate fuels will be reduced. Yet the steady depletion of conventional fuels, though slowed by efficiency improvements, will eventually require the development of alternative fuels.

To meet that need, there are many supply alternatives. Each fuel form (liquids, gases, electricity, or solids) has many substitute technologies already commercially available (but in most cases uneconomical) to replace conventional supplies. Competition among energy-saving technologies, conventional fuels, and alternate fuels for a share of energy service markets has been surprisingly fierce. The unexpectedly large response of both energy savings and conventional energy production to price increases has squeezed alternate fuels out of most energy markets. Yet these alternatives represent an important source of energy abundance that could fuel our energy system for years to come.

[16] Stobaugh and Yergin, among others, have suggested this economic ranking among marginal sources: efficiency improvements cheapest, then alternative technologies; see R. Stobaugh and R. Yergin (eds.), *Energy Futures Report of the Energy Project*, Random House, New York, 1979.

7

Two Abundant Energy Futures

Because of increased efficiency, either outcome results in a more comfortable adaptation to the stresses of the energy transition.

Earlier chapters have chronicled the ways that individuals and companies are reducing the cost of their energy services. When these separate energy-saving decisions are combined, what do they suggest for the future? Specifically, what are the implications of this Least-Cost Energy Strategy for the United States? Will the country be better off than if past practices were continued? To address these questions, we prepared two different projections of U.S. energy consumption to the year 2000:

- A *Business-As-Usual* case, where historical patterns of energy investment behavior continue

- A *Least-Cost* case, where all cost-effective end-use energy investments are realized

Except for its emphasis on energy services rather than energy units, the Business-As-Usual case was intended to produce a result that is similar in nature to conventional energy forecasts. If the Least-Cost outcome is the guide, energy consumers have historically underinvested in energy consumption. Therefore, the Business-As-Usual case assumes that historical end-use investment patterns will continue in the future. In the Least-Cost case, these historical relationships are broken. Energy consumers are assumed to follow Least-Cost guidelines, making future investment and consumption decisions that minimize the cost of energy services. Using these two different criteria for future energy behavior, we forecasted future fuel mix and energy investment patterns using detailed models of end-use energy demand and assumptions about the U.S. economy and world oil prices.[1] While some of these assumptions may prove false—making either forecast an inaccurate prediction of what the future will hold—the results still serve as a useful guide to where the major Least-Cost opportunities lie.

THE LEAST-COST PROJECTION

Chapters 3, 4, and 5 outlined investments that are now occurring in transportation, buildings, and industry, respectively. If all the cost-effective investments outlined in these chapters extend to the nation as a whole, Figure 7.1 shows that in the Least-Cost case, the total U.S. demand for fuels would remain almost level over the next 2 decades.

[1] The analysis described was performed in October 1982 using versions of the Buildings Energy Conservation Optimization Model (BECOM), Industrial Sector Technology Use Model (ISTUM), and the Transportation Energy Consumption Model (TEC) sponsored or developed at the Energy Productivity Center and maintained and operated by Applied Energy Services, Inc. The models represent future end-use energy behavior according to a cost-of-services approach, and included the following key energy and economic assumptions:

1. GNP was assumed to grow at 2.6 percent per year from 1980 to 2000.
2. World oil prices were projected in the near term at about $30 per barrel in 1985 (in 1982 dollars), then rise at 1 to 2 percent per year in real terms to about $50 per barrel in 2000.
3. The difference between the Business-As-Usual and Least-Cost projections was represented by changing the discount rate for energy efficiency investments as follows:

	Business-As-Usual	Least-Cost
Residential buildings	25%/yr	5%/yr
Commercial buildings	15%/yr	5%/yr
Transportation	20%/yr	5%/yr
Industry	10%/yr	5%/yr

4. The analysis includes cost and efficiency estimates of proven technologies. If technology breakthroughs occur, the projections would need to be changed accordingly.

Detailed documentation of the assumptions and results for these projections is contained in *The Least-Cost Update*, Applied Energy Services, Inc., 1925 N. Lynn Street, Suite 1200, Arlington, VA 22209.

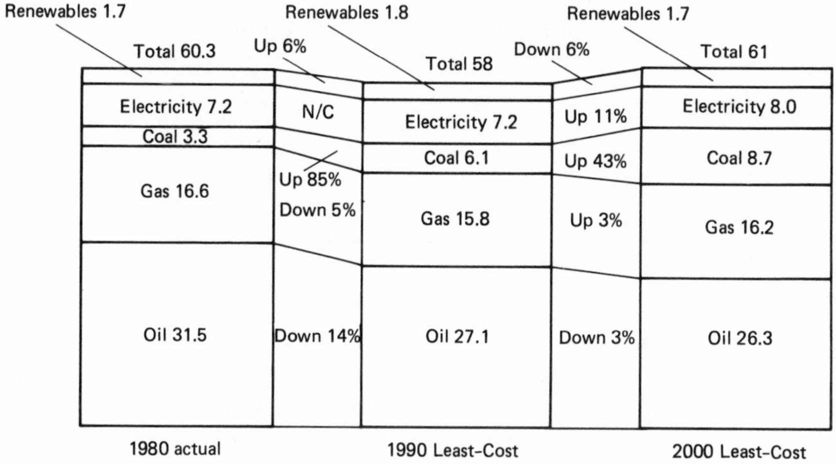

Figure 7.1   Least-cost end-use energy consumption (quads per year).

Demand for fuels would remain flat even though the demand for energy services grew at approximately 2 percent per year over the period and GNP grew at 2.6 percent per year. Even with recent declines in oil prices, projected fuel prices in general continue to make energy efficiency investments a bargain, and these efficiency improvements would continue to displace large quantities of fuel in the future. Without these improvements, fuel use would be 30 percent higher in 2000.

The Least-Cost projections show that end-use demand would actually decline below 1980 levels in 1990 and be about the same as 1980 consumption in the year 2000. Least-Cost demand will decline in 1990 primarily because of the large efficiency savings in existing buildings, which, if retrofitted with Least-Cost conversion devices and shell efficiency improvements, will reduce fuel use in buildings by 25 percent. These savings are large enough to offset increases in the industrial sector, where fuel use is projected to grow by 16 percent despite Least-Cost efficiency improvements.

While savings would continue to grow beyond 1990 (owing to higher energy prices and turnover of end-use capital), the gains in efficiency would not fully offset projected growth in service demand from 1990 to the year 2000. Over the whole period, however, gains in energy efficiency and growth in service demand would tend to offset each other in the Least-Cost projections, leaving fuel use at 61 quads per year in 2000 (about the same as in 1980).

While total end-use demand remains relatively constant in the Least-Cost projections, fuel market shares are projected to change considerably. Oil is projected to compete poorly in future end-use markets. Oil

TABLE 7.1   Percent of Least-Cost End-Use Market Share

Fuel	1980	1990	2000
Oil	52	47	43
Gas	28	27	27
Electricity	12	13	13
Coal	5	10	14
Renewables	3	3	3

SOURCE: Applied Energy Services, Inc., Least-Cost projections.

consumption will decline by 17 percent by 2000, and oil's share of end-use demand will decline steadily, as cheaper fuels and efficiency improvements capture market share from oil products (see Table 7.1). Electricity, gas, and renewables will maintain roughly their current market share, and coal will gain market share in the industrial sector, given Least-Cost assumptions.

Two fuels—electricity and renewables—fare relatively poorly in a Least-Cost forecast compared to current expectations. Centrally generated electricity might grow at only 0.5 percent per year from 1980 to 2000 in a Least-Cost world, well below the utilities' expectations of 2 to 3 percent per year growth. Electric energy efficiency improvements, such as cogeneration, improved lighting, or heat pumps, and lower-than-expected alternative energy prices make it difficult for electricity to gain market share in this projection. Also, renewable energy sources are not Least-Cost unless their costs drop dramatically—a situation which we cannot assume here.

THE BUSINESS-AS-USUAL CASE

To test just how important our Least-Cost assumptions are, we reran our end-use models with an entirely different assumption about consumer behavior—the Business-As-Usual (BAU) case, in which energy consumers would continue to require much higher returns on energy efficiency investments than the organizations supplying fuel and electricity and much higher returns than their costs for obtaining capital.

There are many reasons that consumers have historically invested as if they had higher-than-normal return on investment requirements, and these reasons vary for investors in different end-use sectors. In general, capital has not been readily available to individuals or small firms making energy efficiency investments. More importantly, households and commercial and industrial firms have had competing uses for capital which can make their hurdle rates significantly higher for energy investments. For instance, a home owner may prefer to build an addition to his house or purchase a swimming pool rather than increase the insulation in his attic, unless the anticipated return from energy savings is very high. Similarly, an industrial manager might choose an investment that directly increases profit by increasing sales over an energy efficiency improvement. Consumers seldom base their automobile purchase decisions only on life-cycle costs. Each of these factors impedes energy efficiency investments and raises the effective or observed discount rate—the hurdle rate used to annualize capital or start-up costs over the lifetime of the investment—for end-use investment decisions.

A number of studies have been conducted to determine discount rates (or rate-of-return requirements) that explain the historical pattern of energy investments. Generally, those investors with easy access to the financial markets, good energy information, and a relatively sophisticated knowledge of how to make economic calculations—i.e., businesses—have had the lowest historical discount rates. Households and especially renters have tended to behave as if they faced higher discount rates. For households, high-income families have lower discount rates than low-income households; this factor alone could make a household's real discount rate vary from 5 to 50 percent. For the energy suppliers, we concluded that their investment criteria have been equal to their cost of capital—about 5 percentage points above the underlying rate of inflation, or about 12 percent after tax, assuming an inflation rate of 7 percent.[2] In contrast, we assumed investment premiums of 5 percent for industrial investments, 10 percent for commercial buildings, 15 percent for transportation, and 20 percent for residential buildings in the Business-As-Usual case. These premiums were estimated from historical

[2] A 5 percent real discount rate is equivalent to a 12 percent after-tax cost of capital with the following assumptions:

a. Real discount rate	=	5 %
b. Inflation rate	=	7 %
c. Cost of debt (a + b)	=	12%
d. Cost of equity	=	18%
(a + b + risk premium)		
e. Tax rate	=	50%
f. Debt/equity	=	50%

The after-tax cost of capital is calculated as:

Cost of debt	$= c \times e \times f$	=	.03
Cost of equity	$= d(1-f)$	=	.09
Total cost of capital			$= .12$

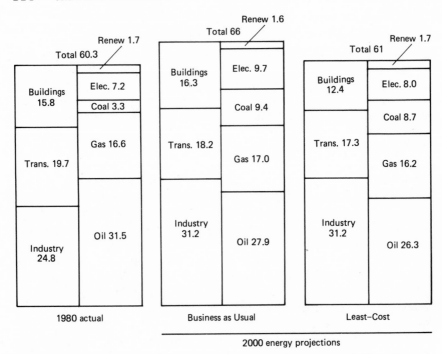

Figure 7.2 Business-As-Usual and Least-Cost projections (quadrillion Btu per year).

behavior in each end-use sector, according to the findings of several studies on this subject.[3] Even though the anecdotes in previous chapters argue that such disparities cannot exist for long, since entrepreneurs will take advantage of such disparities and earn those returns themselves, we assumed in the BAU case that such entrepreneurs would be extremely reticent over the next 20 years.

The results of this exercise were fascinating (Figure 7.2 compares the projected consumption of energy in the Business-As-Usual and Least-Cost cases). Total end-use demand for energy was only projected to be about 10 percent lower in the year 2000 with Least-Cost investments.

---

[3] See, for example, D. O'Neal, K. Corum, and J. Jones, "An Estimate of Consumer's Discount Rates Implicit in Single-Family Housing Construction Practices," *Report No. ORNL/CON-62 of the Oak Ridge National Laboratory*, Oak Ridge, April 1981; J. Hausman, "Individual Discount Rates and the Purchase of Energy-Using Durables," *Bell Journal of Economics*, vol. 10, 1980, pp. 33–54; Henry E. Cold and Robert E. Fullen, *Residential Energy Decision Making: An Overview with Emphasis on Individual Discount Rates and Responsiveness of Household Income and Prices*, Hittman Associates, August 1981, a report prepared for the Department of Energy; Resource Planning Associates, Inc., *The Potential for Industrial Cogeneration Development by 1990*, July 31, 1981, pp. D–24.

End-use demand remained flat at about 60 quads per year in the Least-Cost projection, compared to a growth rate of about 0.5 percent per year in the BAU case. The total cost of energy services was projected to be about 10 percent higher in the BAU case.

Figure 7.2 shows that of all the end-use sectors, buildings energy consumption would be affected the most by Least-Cost decision making. Fuel use in buildings might be 25 percent lower in the year 2000 if building owners take advantage of all the cost-effective energy-saving opportunities in buildings. Most of the difference is in old residential buildings, which are not likely to be retrofitted if historical behavior is any clue to the future.

The extra end-use investments projected in the Least-Cost case have their largest impact on future oil, gas, and electricity consumption (see Figure 7.2). In buildings, oil use would decrease by 30 percent and gas and electricity use by over 20 percent each if Least-Cost investments were made. In the industrial sector, gas would replace coal under boilers and in cogeneration applications (with Least-Cost fuel use, gas consumption "moves" from buildings to industry, and both total gas consumption and gas prices are lower). The extra cogeneration in the Least-Cost projections also helps to decrease central electricity consumption from 1.5 percent per year growth in the BAU case to 0.5 percent per year overall to the year 2000.

The Least-Cost projection also showed an important side-benefit in addition to higher energy savings: energy prices were lower. The net effect on fuel prices is summarized in Table 7.2. The extra energy efficiency investments in the Least-Cost case lower the projected demand for fuels. With lower fuel demand, each fuel market reaches an equilibrium with lower prices. Oil prices might be 10 percent lower and

TABLE 7.2 Business-As-Usual & Least-Cost Energy Prices in the Year 2000

	1982 $/Btu		
	BAU	Least-Cost	% Decrease
Crude oil	10.34	9.14	−12
Natural gas (wellhead)	7.51	5.85	−22
Industrial prices			
Oil	11.82	10.62	−10
Gas	8.55	6.91	−19
Coal	2.71	2.65	−2
Electricity	15.19	14.94	−2

SOURCE: Applied Energy Services, Inc., Least-Cost projections.

gas prices more than 20 percent lower by 2000 with Least-Cost investments.

The Least-Cost investments would lead to significant reductions in the average cost of future energy services. Table 7.3 shows that the cost of energy services might at worst remain constant and at best decline 10 percent relative to GNP in future years. Even in the BAU case, energy efficiency improvements are likely to offset future price increases. With Least-Cost investments, the projected fuel bill would become a smaller share of total energy service costs. For example, in 1980 fuel costs were 64 percent of total energy service costs. With Least-Cost, fuel costs might be reduced to 57 percent of total costs in the year 2000, despite a 60 percent increase in crude oil prices and a threefold increase in the wellhead gas price.

Annual capital costs for end-use efficiency improvements are 7 percent higher in the Least-Cost case, with the largest potential difference—23 percent in 2000—in building investments. But fuel prices and fuel demand are projected to be lower in the Least-Cost case, leading to a potential decrease in the average fuel bill of as much as 18 percent by the year 2000. The net effect is a 10 percent decrease in the total cost of energy services by the year 2000, and if our assumptions are correct, this approximates the lowest cost that consumers might pay for energy services in 2000.

The trend of energy services is projected in Table 7.3 to be different among the three end-use sectors. Energy service costs (as a percent of GNP) are projected to decline in buildings and transportation, as cost-effective investments are likely to offset fuel price increases in both the

TABLE 7.3 Future Cost of Energy Services (% of GNP)

	1980	2000	
		Business-As-Usual	Least-Cost
By Category			
Capital	7.9	8.1	8.7
Fuel	14.1	13.8	11.3
Total	22.0	21.9	20.0
By sector			
Buildings	6.6	6.4	5.7
Industry	4.6	5.7	5.1
Transportation	10.8	9.8	9.2
Total	22.0	21.9	20.0

SOURCE: Applied Energy Services, Inc., Least-Cost projections.

BAU and the Least-Cost cases. In industry, where there is generally less potential for fuel-cutting measures, energy service costs increase somewhat.

COMPARISON WITH GOVERNMENT FORECASTS

As shown in Figure 7.3 the projections of fuel consumption in both the Least-Cost and BAU cases are much lower than comparable government projections (we compared our projections with those of the Energy Information Administration (EIA) at the Department of Energy).[4] All of the projections shown were made with roughly the same assumptions for economic growth (about 2.5 percent per year from 1980 to 2000). However, the projections we made resulted in much lower energy prices than EIA projections. For example, the EIA projected oil prices that reached $88 per barrel in 2000, compared to $50 to $60 per barrel obtained in our projections. Because of greater responsiveness of energy demand to energy prices in our projections, energy prices did not have to rise as much to balance supply and demand.

Even with BAU assumptions about consumer behavior, our projections show significantly lower fuel consumption than EIA. EIA projected that fuel consumption will grow to about 104 quads in the year 2000, a growth rate of 1.5 percent per year. In contrast, the BAU projection shows fuel consumption growing at about half that rate. This difference is significant; the projections we made suggest that considerably more end-use energy efficiency improvements are likely than forecasted by EIA even if current trends continue. Although a continuation of current trends would not tap all the efficiency improvements projected to be Least-Cost, the projection of the amount of efficiency improvements that are likely to occur considerably exceeds those projected by conventional forecasts such as that shown here for EIA.

Perhaps the most important discrepancy is in projected electricity use. EIA projects centrally generated electricity consumption to increase at 2.5 percent per year on average over the next 2 decades. Our results suggest that this estimate is too high; even with BAU assumptions about end-use behavior, centrally generated electricity consumption will grow only at about 1.5 percent per year (and with Least-Cost, consumption of electricity from central-station power plants will grow at only about 0.5 percent per year). Most of the difference is in industrial electricity demands, where our forecasts indicate a substantial reduction in both

---

[4] U.S. Department of Energy, Energy Information Administration, *1981 Annual Report to Congress*, vol. 3, February 1982.

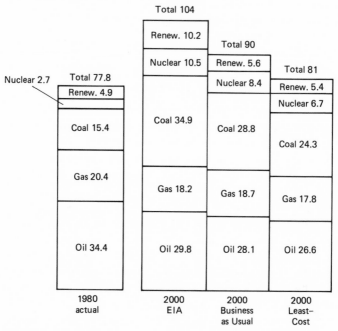

Figure 7.3 Primary energy consumption (quads per year). (*Source*: Applied Energy Services, Inc., Least-Cost projections.)

service demand growth and electricity needs compared to EIA. The BAU projections also shows lower use of renewables, oil, and coal than EIA. Only gas use is projected to be higher.

The Least-Cost results indicate that a substantial amount of additional cost-effective efficiency improvements remain to be tapped by investing in energy productivity. These potential investments in buildings, industry, and transportation could total as much as $1.7 trillion (in 1982 dollars) over the next 2 decades. Table 7.4 summarizes the most productive technologies ranked by investment potential within each end-use sector. The major categories in each sector are surprisingly unsophisticated: auto fuel economy improvements, which require little advanced technology; building-shell improvements, which use mundane materials and devices; and industrial boilers and furnaces, whose basic design has not changed much for several decades. Other large categories of potential cost-effective energy investments include efficient lighting and appliances, commercial HVAC equipment, appliances, truck fuel economy, and various efficiency improvements in industry.

In the BAU case, fewer end-use investments are made, although the total amount is still quite significant—$1.3 trillion total investment

TABLE 7.4 Projected Investments in Energy Efficiency Improvements

Technology group	1980–2000 Cumulative Investments (billion 1982 dollars)		
	Business-As-Usual	Least-Cost	Difference
Buildings			
Shell improvements	36	316	+280
Thermal devices, lighting, and appliances	216	216	
Conventional HVAC equipment	241	144	−97
Heat pumps and pulse-combustion furnaces	9	99	+90
Commercial buildings cogeneration	0	19	+19
Subtotal	502	794	+292
Industry			
Boilers and furnaces	142	127	−15
Efficiency improvements	64	63	−1
Cogeneration technologies	9	41	+32
Self-generation technologies	35	31	−4
Subtotal	250	262	+12
Transportation			
Auto fuel economy	292	298	+6
Light and heavy truck fuel economy	179	206	+27
Rail electrification and miscellaneous improvements	64	64	
Aircraft improvements	51	53	+2
Subtotal	586	621	+35
Total investments	1338	1677	+339

during the period 1980–2000. If Least-Cost results continue, investments in building-shell improvements will increase considerably (especially in retrofits of existing buildings), and a shift will occur from less-expensive conventional heating and cooling equipment (such as oil, gas, or electric resistance heat) to high-efficiency conversion devices (such as heat pumps or gas pulse-combustion furnaces). In industry, the major difference is the amount of expected investment in cogeneration. Because cogeneration technologies are capital-intensive, they become more attractive when lower rates of return on investment are required before a project goes forward. In the transportation sector, fuel economy investments are slightly higher in the Least-Cost case.

The absolute amount as well as the difference between these two projections highlights the remaining investment opportunities in energy productivity technologies. If current investment trends continue (the BAU case), there would likely be a market of about $350 billion of unrealized cost-effective investments over the next 2 decades. Building-shell improvements are by far the largest single market, followed by high-efficiency heating and cooling equipment (heat pumps and gas pulse-combustion furnaces), cogeneration, and fuel economy improvements. These end-use investment markets are so large that small fractions of the smallest category could launch a successful new company. Large fractions could transfer cash-rich companies into new high-growth markets for the 1980s.

As we have seen previously, the emerging concept of an energy service company, a third-party company which takes advantage of the potential investments in end-use technologies depicted in the Least-Cost projection, is offering a whole new approach to tap these end-use markets. These companies sell energy services (for example, steam or heating and lighting) at competitive prices using new, more efficient technologies and requiring returns for their investment that are typical of companies supplying fuel.

Either of these energy futures—what we call the Business-As-Usual or Least-Cost projections—show a surprising amount of energy abundance over the coming decades. Even historical investment patterns (the BAU case) lead to less energy use and lower energy prices than conventional forecasts indicate. The Least-Cost outcome, where consumers break from the past and make only cost-effective energy investments, results in even lower energy service costs. Because of increased efficiency, either outcome results in a more comfortable adaptation to the stresses of the energy transition that have created a new energy reality.

Furthermore, the Least-Cost case presents an entrepreneurial opportunity akin in size to semiconductors or genetic medicine. The innovations occurring have more to do with marketing and management than with science, but are nevertheless exciting. Perhaps we should only take this as confirmation of the adage that problems create opportunities, but we prefer to see it as a demonstration of the incomparable strength and diversity of American consumers.

Of course, other outcomes besides the two futures shown here are possible—futures assuming lower or higher oil prices, less or more economic growth, or other potential planning contingencies. Even in such cases, however, we were unable to come up with a plausible set of circumstances in which a comfortable accommodation is not likely. Energy abundance appears to be a condition Americans are creating and will continue to create, not something requiring a turnaround in our actions, our thinking, or our laws.

8

Some Myths About Energy

Contrary to much conventional wisdom, using more energy on the one hand, or sacrificing comfort and economic growth on the other, are not the only alternatives open to the American people.

If the results described in Chapter 7 are any indication, there is an unnecessary tendency to hang onto and to believe the apocryphal energy stories of the early 1970s. In spite of the buildup of evidence to the contrary, many energy myths live on. These myths are continuing to lead us on some very unproductive ventures, and most still form the foundation for U.S. energy policy. In this chapter we would like to take a shot at destroying the major myths that remain.

MYTH 1: ENERGY IS SCARCE

As we indicated previously, underlying most of the nation's energy-related policies and actions is the dominant belief that "energy is scarce." Historically, the fear of running out has motivated much of the nation's energy strategy. In 1974, two of us helped support that view by posting large numbers in our Federal Energy Administration windows to remind the public how rapidly petroleum reserves were running out and to justify our conservation initiatives—an approach we now gag on.

After the embargo—under pressure from voters to act in response to these scarcity messages—Congress passed laws to allocate "shortages" and to control energy prices in an equitable manner. For example:

- A program was established requiring certain utility and industrial operations to switch from scarce oil and gas to coal.

- The Fuel Use Act legislated many complex rules restricting natural gas and oil use in industry and utilities so that dwindling supplies would be available for use in homes and other "priority" areas.

- Major tax credits and other stimulants were created to encourage conservation or conversion to renewable technologies which tapped the infinite and inexhaustible energy of the sun.

- Standards were implemented or proposed for the energy efficiency of automobiles, buildings, and appliances.

- Grants and information designed to facilitate conservation were introduced.

- The Synthetic Fuels Corporation was created in an attempt to accelerate the development of liquid and gaseous fuels and prepare for the dwindling of conventional oil and gas supplies

None of these programs, in hindsight, looks very smart. Nevertheless, worried that support for this legislation was beginning to wane, supporters have launched a new campaign. As recently as the fall of 1982, the International Energy Agency (IEA) published a report claiming, in effect, that another crisis was just around the corner. The IEA warned energy users and policy makers not to be complacent during the present oil glut. The combination of growing oil demand later in this decade and shrinking oil production "could again set the stage for significant oil price increases and major market disruptions."[1] A *Washington Post* editorial commended the IEA, but *The Wall Street Journal* wondered whether the IEA would ever learn about the workings of the market. "With oil price decontrol and a declining rate of dollar inflation, the energy crisis is over. But the melody lingers on—or at least the IEA has been able to stir a few headlines with a report that the crisis lurks just a few years around the corner. Sometimes we wonder whether anything has been learned."[2] Of course, *The Wall Street Journal* did not say that they themselves had consistently underestimated the demand response to price previously, preferring instead to emphasize supply elasticity.

Oil is not the only energy commodity that elicits fears of scarcity. Many senior officers of electric utilities fear that a prolonged surge in

---

[1] International Energy Agency, *World Energy Outlook*, 1982, p. 9.
[2] *The Wall Street Journal*, Oct. 22, 1982.

economic growth over the next few decades will create a scarcity of electric generation capacity in the 1990s. In 1983, Warren Anderson, the chairman of Union Carbide Corporation, said: "The recent easing of demand for electricity, combined with increased supplies of petroleum products, do not signal an end to America's energy problems. Both conditions are in large measure caused by severe recession. A strong upturn in industrial activity will quickly reverse these conditions."[3] Shortfalls of natural gas are also feared: In 1983 a senior gas utility executive warned of a natural gas shortfall in the 1990s because the deferral of projects will lead to a lack of supply.[4]

Since fossil fuels are indeed finite, and since we are in the midst of a unique energy transition, there have been few questions about the validity of the scarcity rationale for U.S. and foreign actions. Yet the evidence in previous chapters shows that there is not now, nor is there likely to be, a scarcity of ways to provide this nation with the energy *services* it needs. What is apparent is that energy commodities are only one means of supplying the needs that are vital to our economy, namely services such as heating, lighting, cooling, and mobility. As a demonstration, the U.S. economy in 1983 used about 23 percent less energy to produce each dollar of product than in 1973.[5]

Our projections, described in Chapter 7, revealed that by the year 2000, each dollar of GNP would require about 60 percent of the energy used in 1980 if Least-Cost energy systems were utilized. Economic growth, projected at about 2.6 percent per year through the end of the century, can occur without any increase in energy consumption through greater investment in more efficient end-use equipment and materials—which in turn makes fuel available in sufficient amounts to satisfy growth in energy service demand.

The conclusion we have reached is that while a scarcity of oil and gas is at the root of the present energy transition, there is no scarcity of ways to provide energy services. An abundance of options are available to meet America's energy service needs; several solutions besides the ones we arrived at are possible. Not only are substitutes for current energy sources technically available, many of them are less costly than oil, natural gas, and other fuels that have been considered essential and irreplaceable. Contrary to much conventional wisdom, using more energy or sacrificing comfort and economic growth are not the only alternatives available to the American people. Neither are the options

---

[3] Warren Anderson, Chairman of Union Carbide Corporation, *The Wall Street Journal*, Apr. 7, 1983, p. 15.

[4] "U.S. Seen Facing Natural Gas Shortfall in 1990s," *Oil and Gas Journal*, vol. 81, no. 10, Mar. 7, 1983, p. 55.

[5] Washington Analysis Center, Monthly Statistical Report, Oct. 31, 1983.

simply choices between oil, gas, coal, solar, or nuclear. As the nation shifts from fears of scarcity to an energy services paradigm, it is becoming apparent that the total amount of fuel used will be much lower than traditional energy forecasts have projected. Because hundreds of nonfuel technologies can help deliver the desired services at lower costs, growing human needs and wants can and are being met with very modest increases in fuel use. All of these options for minimizing the cost of energy services are helping to make the current energy transition much less traumatic than expected and are working to dispel the popular conception that energy is scarce.

MYTH 2: HIGHER PRICES ARE INEVITABLE

Second on the list of fundamental energy beliefs is the oft-repeated warning that energy prices will continue to rise as they have in the recent past and that there's little that consumers can do in response. As a result, we expect that the costs of heating, cooling, lighting, and driving will rise faster than anything else for the foreseeable future. To quote Samuel Dix, a congressional witness, "Higher energy costs and a lesser quantity of available energy will affect the total economy, but the effect will be most severe on the individual as a consumer. . . . Every man cannot expect to live as a king, lighting a thousand candles and maintaining a castle to entertain his friends."[6] While economists have traditionally emphasized that rising prices will increase supplies and reduce demand, this fear of energy price increases was responsible for establishing oil price controls and the inordinate difficulty in finally getting them removed. The same basic fear underlies the current resistance to decontrolling natural gas prices and free-market electric rates.

The assertion that energy costs will continue to rise more rapidly than other items has been difficult to refute. The pain of paying fuel and electric bills for the last 10 years has led to the belief that there is little or no competition in the energy sector. But the time has come where this concern ought to be questioned. The Least-Cost results support current experience that there is a limit to such price rises. As the recent decline in world oil prices has illustrated, energy prices have now risen to such levels that ample alternatives exist to keep these costs in check. Competition is now rampant, not just among energy sources but also between fuels and end-use technologies like efficient appliances and light bulbs, and high mileage cars.

[6] Samuel M. Dix, *Hearing before the Subcommittee on Energy and Power of the Committee on Interstate and Foreign Commerce*, "A Critical Decision for the United States Economy," House of Representatives, Mar. 23, 1977, Washington, D.C., G.P.O., Serial No. 95-15, 1977, pp. 227–231.

TABLE 8.1 Historical and Projected Energy Price Increases

	% per year	
	1973–1980	1980–2000
Crude oil	18	2–3
Wellhead gas	23	5–7
Gasoline	9	1–2
Distillate oil	19	2–3
Gas	14	3–4
Coal	8	2
Electricity	5	1
Energy service costs	+5.5	−0.5

SOURCES: Historical data from Energy Information Administration, *State Energy Price System: Overview and Technical Documentation*, November 1982, vol 1; *Monthly Energy Review*, January 1983 and *Annual Report to Congress*, 1981. Fuel price and energy service cost projections from AES, *The Least-Cost Update*, op. cit. (range of BAU and Least-Cost projections).

As Table 8.1 shows, the rapid price increase and escalating cost of energy services of the past decade may be over. Future energy price increases are likely to be much slower than during the 1973–1980 period, and energy service costs are likely to decline. By choosing a less-expensive mix of fuels and energy-using devices, the nation has begun a trend to reduce the cost of energy services per dollar of GNP, a trend that should continue throughout the remainder of this century. In that event the dollar amount spent for fuel will be significantly reduced—even with higher oil, gas, and elecricity prices—and its share of energy services costs replaced by more efficient energy-using capital equipment.

MYTH 3: ENVIRONMENTALISTS ARE CAUSING THE ENERGY PROBLEM

One of the most popular scapegoats of the energy crisis has been the environmental community. "Environmentalists are causing the problem" has been a major premise of many business and political leaders. For example, the automobile pollution standards were attacked from all sides as contributors to energy inefficiency, especially in the early years of energy awareness. President Ford's 1975 proclamation epitomized this thinking: "My National Energy Conservation Plan will urge Congress to grant a five year delay on higher automobile pollution standards in order to achieve a 40 percent improvement in miles per gallon."[7] Later, the two turned out to be essentially independent of each other.

[7] Gerald R. Ford, "Address by the President on Live Television and Radio," Jan. 13, 1975, *Executive Energy Documents No. 95-144*, Washington, D.C., 1978, p. 172.

Environmental standards for mining and burning coal have been cited repeatedly as causes for inadequate energy supplies.[8] The Reagan administration's attempt to unnecessarily expand oil, gas, and coal leasing on environmentally sensitive federal lands over environmentalist protests is a continuation of this old but apparently irrelevant argument.

The Least-Cost analysis is not intended to be a study of ways to resolve energy-environmental conflicts. However, in the results we tabulated, we required that any energy service technologies—either on the end-use or supply side—had to meet environmental standards in effect at the time in order to be considered. Hence, the choice of energy technologies is based on the total cost of energy services to consumers, *inclusive* of the cost of meeting current environmental requirements. For example, the automobile fuel economy improvements we used include engine, drive-train, and vehicle technologies that comply with the present version of the Clean Air Act. Similarly, the projected major increase in coal use in industry required a mix of fluidized bed boilers, low Btu coal gasification, and conventional coal-burning equipment with suitable emission control devices to be consistent with environmental standards.

The Least-Cost results also projected the increased use of clean-burning gas in industry, both for combustion applications and for the generation of electricity at industrial sites—some of which can be exported to the electricity grid. This latter trend, in conjunction with implementation of efficiency improvement technologies in buildings and industry, will serve to reduce the use of environmentally sensitive coal and nuclear electric generation far below levels traditionally projected. The coal and nuclear plants that are built will serve mainly to replace more expensive oil and gas power stations, since the overall increase in central-station generation over the 2 decades is projected to be only 0.5 percent. The reduction in the coal and nuclear growth rate because of its lack of economic competitiveness compared to end-use technologies and smaller, decentralized generating plants significantly reduces the much-discussed environmental risks from these technologies.

Also in these results, synthetic liquid fuel supplies were determined to be a marginal source—less than 200,000 barrels per day—entering the fuel mix late in the century. Larger quantities of production have raised environmental issues such as land and water use in the west. But cost-

[8] For example, energy secretary James B. Edwards in his confirmation hearing at the beginning of the Reagan administration said, "We have an awful lot of need to use coal . . . we need to allow our people to mine coal and the restrictions are pretty tough on the miners . . . we need to burn the coal and we need to allow our people to burn it a little more freely and a little more easily without the absolute restrictions that they have to work under today." James B. Edwards, *Nomination Hearing before the Committee on Energy and Natural Resources*, U.S. Government Printing Office, Serial No. 97.2, Washington, D.C., 1980, p. 37.

competitive synfuel production will likely be modest enough and sufficiently far into the future that careful project planning and use of best available technologies can meet rigorous environmental requirements.

The Least-Cost study did not deal directly with the environmental cost of increased strip mining, the health effects of expanded deep-coal mining, chemical and thermal water pollution impacts, or any of the emerging environmental questions like the release of CO_2 and acid rain. But again, since the results imply dramatically less coal and nuclear-generated elecricity than government and other projections, their negative environmental impacts would be minimized.[9]

In summary, the research indicates that future energy services required to meet the needs of a growing economy can be met without unraveling environmental regulations. Environmental energy issues need not be argued on the basis of whether environmental standards allow sufficient energy to supply our future needs. The issue is more accurately framed in terms of the value of enhanced environmental quality versus the increase or decrease in the cost of energy services to the consumer if standards are strengthened or relaxed. Whether particular federal off-shore oil tracts or coal-mining lands are opened for exploration, for example, should depend on the perceived value of that land in its present state versus how much the exploration and resource extraction are likely to reduce the total cost of energy services. These equations are usually resolved politically because the values for environmental quality and health effects are usually not factored into the calculations of the marketplace. The context of this debate is being reshaped by the Least-Cost results, which suggest that there is no basis to the argument that environmental standards must be weakened to meet national-security or economic-growth requirements. Simply stated, an economically efficient energy policy will have a benign effect on the environment. Therefore, current environmental regulations are not a significant barrier to meeting energy service needs.

MYTH 4: WE CANNOT CONSERVE OUR WAY TO ECONOMIC GROWTH

A fourth belief has found favor in recent years: "We cannot conserve our way to economic growth."[10] A corollary to this myth is that growth in fuel use—particularly electricity—is essential to our economic health and

---

[9] For an expanded version of this argument see A. Lovins et al., *Least-Cost Energy: Solving the CO_2 Problem*, Brick House Publishing, Andover, 1982.

[10] This statement by energy secretary James B. Edwards, was confirmed and clarified for us on July 1, 1981, by Robert C. Odle, assistant secretary of energy for Congressional,

comfort. This belief stems from persistent statements that conservation means deprivation—turning down thermostats, driving fewer miles, or producing fewer products[11]—and leads to an energy strategy conclusion that large new sources of energy must be found or built or our economy and lifestyle will suffer. It also leads to major efforts to undo environmental regulations and stimulates efforts to push the production of nuclear power and subsidize synthetic fuels and solar development.

While future economic growth will require growth in energy services, it does not follow that economic growth requires additional energy use. Our projections showed that future fuel use will likely remain roughly at current levels while the economy will grow at 2.6 percent per year over the next 2 decades. Furthermore, the long-standing lockstep relationship between electricity growth and GNP growth is likely to change: electricity sales are likely to grow more slowly than previously called for by GNP growth. Both of these results stem from projected gains in energy productivity—ways to increase the benefits obtained from each Btu of fuel. Current trends suggest energy productivity will be a major stimulus to economic growth, not a deterrent. This will come about because the cost of saving energy through improving vehicle mileage and furnace and lighting efficiency is lower than the cost of most sources of energy supply.

It follows that when the total cost of services—both fuel and capital charges—is reduced, economic resources that previously had been used

---

Intergovernmental and Public Affairs. Explaining the contexts in which Secretary Edwards reiterated this position, Odle stated, "The secretary believes that conservation is an important building block of any energy plan, but it alone will not stimulate the kind of economic growth that there must be." While other statements of Secretary Edwards and Secretary Hodel have been slightly more bullish about energy end-use investment, this statement and others like it have tended to obscure the fact that for the foreseeable future productivity is the most cost-effective energy alternative from the point of view of enhancing economic growth.

[11] See, for instance, Nixon, Richard M., "Address on the Energy Emergency," November 7, 1973. Ford, Gerald R., op. cit. Carter, Jimmy E., "Address to the Nation," July 15, 1979, *The Washington Post*, July 16, 1979, p. A14.

Nixon: To be sure that there is enough oil to go around for the entire winter, all over the country, it will be essential for all of us to live and work in lower temperatures. We must ask everyone to lower the thermostat in your home by at least 6 degrees, so that we can achieve a national daytime average of 68 degrees. ... In offices, factories and commercial establishments, we must ask that you achieve the equivalent of a 10-degree reduction by either lowering the thermostat or curtailing working hours.

Ford: There must be volunteer efforts to cut gasoline and other energy use. ... This plan requires personal sacrifice.

Carter: And I am asking you—for your good and your nation's security—to take no unnecessary trips, to use carpools or public transportation wherever you can, to park your car one extra day per week, to obey the speed limit and to set your thermostats to save fuel.

TABLE 8.2 Average Energy Investments (in billion 1982 dollars)

	Estimated 1980	1980s Least-Cost	1980s Business-As-Usual	1990s Least-Cost	1990s Business-As-Usual
Supply					
Oil imports	93	58	68	59	77
Oil and gas (domestic)	61	76	79	102	106
Electric utilities	40	12	28	17	23
Coal and nuclear fuel	8	10	11	12	15
Total	202	156	186	190	221
Demand					
Buildings	26	55	29	25	24
Industry	8	12	11	12	13
Transportation	16	32	30	30	29
Total	50	99	70	67	66
Total	252	255	256	257	287
Percentage of average GNP	8	7	7	5	6

SOURCE: Historical estimates for oil and gas (domestic) and coal-nuclear fuel from Bankers Trust Company, *U.S. Energy and Capital, A Forecast 1980–1990*, New York, 1980. Electric utilities capacity and facilities under construction based on data from North American Electric Reliability Council, *Electric Power Supply and Demand 1982–1991: Annual Data Summary Report for the Regional Reliability Councils of NERC*, Princeton, 1982, and U.S. Department of Energy, *Capacity Utilization and Fuel Consumption in the Electric Power Industry, 1970–1980*, July 1982. Investment forecasts are derived from the AES Least-Cost projections.

to provide energy services are liberated for investments elsewhere in the economy—a process that creates additional economic growth. Therefore, what is important is not whether the United States can conserve or produce its way to economic growth but what combination of each minimizes costs and releases the most resources to other productive uses outside the energy sector. Table 8.2 shows that if either the Least-Cost or Business-As-Usual approach continues, the total energy sector investment in the 1990s would be only modestly higher than the estimated 1980 investment rate. Total energy sector investment would actually drop as a share of GNP, improving the total productivity of the economy.

Strictly speaking, Table 8.2 is not concerned solely with investment, since imports represent an operating expense. Rather, this table represents a hybrid of different sectors of the national income accounts to test the degree to which enhanced investment in energy productivity might strain domestic capital availability. In this context, we view the energy system as providing energy services to domestic production, nonenergy domestic consumption, investment goods production, government operations, and nonenergy net exports. The "external" inputs to this system are domestic energy investment and energy imports. Whatever resources can be drawn from these sources for cost-effective investment will represent transfers within the energy system and cause no incremental demand on capital markets.

It is evident that the nation can conserve its way to economic growth if energy conservation and energy productivity mean the same thing, and we think they do. Investing in energy productivity means following an economic growth strategy that opens new markets, reduces the cost of providing energy services to the consumer, and maintains the level and quality of services Americans desire.

MYTH 5: OIL IMPORTS ARE THE PROBLEM

One of the most popular explanations of our energy problem is that "our real problem is our dependence on foreign oil." This belief was the rationale behind President Nixon's Project Independence in November 1973 and for many of the policies that grew out of that initiative. Similarily, the primary goal of both President Ford's and President Carter's national energy programs was to restrict imports.[12] The policy objective of finding something to replace imports, accompanied by the assumption that energy in general was scarce, led to mandatory coal-switching rules, mandatory automobile, building, and appliance performance standards, substantial subsidies for solar, conservation, nuclear, and synthetic fuels, and continuation of many policies subsidizing development of conventional fuel forms.

Against the background of Arab embargoes, the Iranian revolution, and gasoline lines, it has been easy to blame our problems on foreign oil. Yet we end up believing that identifying dependence on foreign oil as the principal energy problem has been overstated. It appears that the nation's dependence on foreign oil is merely a symptom—a symptom of

---

[12] According to *The National Energy Plan*, (Executive Office of The President, Energy Policy and Planning, 1977, page IX), U.S. energy policy has three overriding energy objectives:

As an immediate objective that will become even more important in the future, to reduce dependency on foreign oil and vulnerability to supply interruptions

In the medium term, to keep U.S. imports sufficiently low to weather the period when world oil production approaches its capacity limitation

In the long term, to have renewable and essentially inexhaustible sources of energy for sustained economic growth

a much broader energy market imbalance. Ironically, our dependence, even though declining, is partially the unintended result of the various political efforts during this transition period to reduce imports. One example of this was limiting the use of gas under industrial and utility boilers, causing a continuation of fuel oil use. With the benefit of hindsight, had these efforts focused more broadly on allowing the energy services marketplace to work while simultaneously creating a strategic oil stockpile to reduce our vulnerability in coping with import interruptions, the nation's energy situation would be even more secure today.

Political efforts to reduce imports stemmed from a false belief that oil exporters could charge anything they wanted for their oil because "consumers have no alternatives." Fortunately, it has been demonstrated that there simply is too much competition for the OPEC cartel. They, like everyone else, must not exceed a market clearing price. At this point, energy users are free to choose a mix of fuels and end-use equipment on the basis of lowest cost, letting reduced import levels be the result, not the objective, of the strategy.

Even at the current reduced OPEC oil prices, imported oil is still one of the most expensive means available to supply many of our energy services. If people continue to pursue the objective of reducing the cost of energy services, rather than reducing imports *per se*, there will continue to be many alternatives to imports—so many, in fact, that imports will be greatly reduced as a component of the U.S. energy system by the year 2000. The Least-Cost figures show total imports should be about 4 million barrels per day by 2000 (29 percent of projected U.S. oil consumption), versus the 6.4 million barrels per day (37 percent) experienced in 1980. Furthermore, the Least-Cost figures prepared in 1979 indicated that even at $16 per barrel, roughly half of the 8 million barrels per day then being imported were noncompetitive in the U.S. market.

In 1981 and 1982, with the official OPEC price at $34 per barrel and many producing nations announcing price cuts, newspaper articles were popularly titled "The Decline of the OPEC Cartel,"[13] or "Oil Supplies Depend on the Market, Not on Politics,"[14] or "OPEC's Arrogant Blackmail Changed Disaster into Glut."[15] The reality was setting in. According to a *Wall Street Journal* editorial, "a shift has taken place in oil

[13] Peter F. Drucker, "The Decline of the OPEC Cartel," *The Wall Street Journal*, Nov. 26, 1982.

[14] Fred S. Singer, "Oil Supplies Depend on the Market, Not on Politics", *The Wall Street Journal*, Aug. 27, 1982.

[15] Hobart Rowan, "OPEC's Arrogant Blackmail Changed Disaster into Glut," *The Washington Post*, Apr. 2, 1982.

from a 'sellers' to a 'buyers' market—every time OPEC raises prices another notch, it suffers from a drop in demand and also makes alternative energy sources more price-competitive."[16] Now, these conclusions are even more obvious.

Despite this evidence that oil markets are more responsive to price than expected, the United States and the world outside Communist countries (WOCA) will remain vulnerable to sudden oil interruptions. Measures targeted at reducing that vulnerability, discussed further in Chapter 10, should receive the highest priority. It is also possible that strong world economic growth could tighten oil markets once again. But the belief that imports are the starting point from which all energy policy should flow is a myth, and perpetuating that myth will exacerbate the real problem rather than contributing to its resolution.

MYTH 6: ELECTRICITY IS A NATURAL MONOPOLY

The principal basis for the electrical regulatory system as it exists in the United States today is that the generation of electricity is a natural monopoly. Along with others, we challenged this premise first in 1979.[17] Even then, data indicated that the basis upon which utilities have been regulated was questionable. The suggestion that it might be appropriate to deregulate electricity generation while keeping the transmission and distribution functions as a regulated monopoly was not popular at the time. Our later studies provided additional evidence that the debate should continue, and now, electricity deregulation has become a much-discussed issue.[18]

In recent years it has been concluded that regulating the price of oil and gas below the market is inappropriate because it fails to stimulate sufficient reduction in demand or investment in new supplies. There is also considerable agreement that competition among suppliers will hold prices in check. Our analysis takes those conclusions one step further—

---

[16] See, for example, "Energy Crises Revisited," *The Wall Street Journal*, Oct. 22, 1982, p. 28.

[17] Roger W. Sant, *The Least-Cost Energy Strategy*, Carnegie-Mellon University Press, Pittsburgh, 1979, p. 44.

[18] See, for example, John Bryson, "Remarks to the Commonwealth Club of California," San Francisco, May 15, 1981; Irwin M. Stelzer, "The Electric Utility Deregulation Debate: They Do Not Know What They Believe; They Believe That They Know," paper presented at the EEI Rate Research Committee Meeting, Williamsburg, Virginia, Sept. 21, 1981; and Edwin Berlin, Charles J. Cicchetti, and William J. Gillen, "Restructuring the Electric Power Industry," *Electric Power Reform: The Alternatives for Michigan*, Institute of Science and Technology, University of Michigan, Ann Arbor, 1976.

that the market for electricity production is potentially as competitive as it is for oil and gas. Today, there is obvious competition among the major technologies that generate electricity from central stations—coal, nuclear, oil, gas, and hydro power. If coal and nuclear plants are generally cheaper sources of electricity than oil and gas plants, competitive forces, if allowed to operate, will gradually force oil and gas out of the market. If not, oil and gas should be allowed their fair share. Although we did not analyze in detail the cost differential between oil, gas, coal, and nuclear, the potential competition among the four to supply electricity is very real.

One study characterizes the conventional wisdom among utilities as being "total (nuclear) generation costs will equal or slightly undercut those of new coal plants on a national average basis." Yet it found that the generating costs of new nuclear plants will exceed those of new coal plants by 20 to 25 percent on average.[19] Clearly the coal-nuclear debate is unresolved. In this context, the nuclear option (as well as the abundance of coal) serves to limit the real price of coal. Meanwhile the Gas Research Institute insists that gas-fired electricity generation will continue to be competitive. Unconventional renewable sources of electricity may also become competitive in this market, depending upon technical developments.

Even more pertinent to the debate over competitiveness is the fact that decentralized technologies are now available to produce electricity at costs below that produced from new coal or nuclear plants. Cogeneration and small power plant technologies provide strong competition for traditional utility plant designs. Our analysis included eight different cogeneration systems that are commercially available today to produce electricity and steam (or hot water) on an industrial site. These systems can be fueled with oil, gas, coal, petroleum coke, and wood waste. As described earlier, by establishing cost minimization as the criterion, several of these technologies capture large segments of the industrial market. Other ways of generating electricity could also become competitive—photovoltaic cells, wind generation, minihydroelectric plants, biomass, and fuel cells—if their costs continue to come down.

Are regulated utility monopolies the best way to ensure that the investment will be made in these new less-expensive technologies? Certainly an argument can be made that the delivery of electricity to homes and factories in a particular region should be the responsibility of a single institution (a utility). Regulation of the price that a utility can charge for distributing electricity—essentially a toll charge—and the

---

[19] Charles Komanoff, *Power Plant Cost Escalation*, Komanoff Energy Associates, New York, 1981, p. 2.

return on investment it can earn may be appropriate too. But since the generation of electricity is potentially as competitive as other energy markets, serious consideration should be given to separating generation from distribution. Without minimizing the practical difficulties of accomplishing such a major institutional change, we conclude that the possible economic benefits of increased competition in this sector warrant widespread experimentation.

Today there is great concern about the financial plight of electric utilities. On March 6, 1981, the Federal Energy Regulatory Commission held a hearing on the financial crisis facing the utility industry, at which most industry and Wall Street spokespersons suggested that electrical rates must be increased to give utilities higher rates of return on investment.[20] These higher returns, they argued, are necessary if utilities are to raise capital for new electrical generating capacity. Yet our work suggests that this approach to the financial health of the industry stems from a misunderstanding of the competitive forces now affecting the electric industry. The data indicate that centrally generated electricity *at current prices* is already uncompetitive in several energy service markets. In many cases raising the price will just exacerbate electric utilities' problems. As one of us remarked at that time, "To say that utilities' problems can be solved by raising rates is much like saying that Chrysler's problems can be solved by raising the price of their automobiles."[21]

Utility executives are now recognizing that in the competition to deliver energy services, centrally generated electricity is in many cases losing to conservation technologies ranging from new light bulbs to more efficient electric motors and from appliances and insulation to new industrial processes that minimize electric costs. In most cases, it is also at a significant competitive disadvantage with cogeneration and other decentralized sources of electric generation that are now exempt from regulation.

Obviously, from a utility's point of view this situation suggests a shift away from total reliance on central-station generation of elecricity and its delivery to consumers. The need is to facilitate investments in these technologies to make utilities more competitive in delivering energy services. It is also important to ensure that all technologies for generating electricity are allowed to compete in the marketplace. But as a

---

[20] Dennis Bakke, Gordon Corey, William Grigg, and Joseph Swidler, *Informal Conference on the Financial Condition of the Electric Power Industry*, Federal Energy Regulatory Commission, Washington, D.C., Mar. 6, 1981.

[21] Op. Cit., p. 120.

nation, it is vital to recognize that the generation of electricity is not a natural monopoly. In fact, the "obligation to serve" justification may be even more shortsighted. Electricity is more appropriately considered a commodity, not an entitlement.

MYTH 7: FUEL FOR CARS IS THE PROBLEM

Myth number 7 is that "the most nonreplaceable energy is the liquid fuel required for autos and other transportation." "The U.S. runs on gasoline; liquid fuels are the hub of the complex issues known as the energy problem," according to one prominent expert whose views on this subject are widely respected.[22] This assertion provided the basis for adopting automobile efficiency standards in 1975.

It is easy to become snagged on this myth, since it is so difficult to imagine transportation alternatives to gasoline or diesel fuels. Yet, like the other energy myths, the focus on fuels results in misleading conclusions. There *are* alternatives to liquid fuels in the transportation sector—alternatives that can reduce costs and fuel use. For example, as we showed in Chapter 3, the transportation sector should require 12 percent less fuel in 2000 than today because of significant opportunities to further improve the "mobility" per gallon of fuels used by autos, trucks, and aircraft. Two of our former colleagues indicated in their 1980 study that in the next 10 years an investment of $50 billion by auto companies will produce roughly twice as much "mobility" as would the same investment in synthetic fuels.[23] Furthermore, the cheapest synthetic fuels, shale oil and methanol, have a good chance at becoming competitive with gasoline without government subsidies within the century.

This situation of emerging competitive fuels and better-designed automobiles is enhanced when oil is priced out of the space heating, electric generation, and industrial steam markets and can only compete in the transportation sector. In the Least-Cost projections, oil consumption shifts from the buildings and industrial sectors into transportation. This development, combined with increased fuel economy, permits savings in the transportation sector to back out foreign oil from the

[22] Daniel Yergin, et al., *The Dependence Dilemma*, Center for International Affairs, Harvard University, Massachusetts, 1980, p. 16.

[23] Richard H. Shackson and H. James Leach, *Maintaining Automotive Mobility: Using Fuel Economy and Synthetic Fuels to Compete with OPEC Oil*, Carnegie-Mellon University Press, Pittsburgh, 1980.

world markets, reducing U.S. imports sharply. Thus the challenge in providing adequate liquid fuels for transportation is, in our opinion, overrated—in short, a myth.

MYTH 8: NATURAL GAS PRICES WILL RISE WITH OIL PRICES

Throughout the natural gas debate both opponents and proponents of deregulation have argued that "without controls, natural gas will bring a price equal to or greater than the price of oil." The basis of that argument is that gas can be used to replace oil in many end uses. Since it is cleaner burning than oil, it should sell at a slight premium above oil prices.

This assumption is central to the objections to natural gas deregulation, for it is used to argue that gas prices could rise sharply when deregulation occurs. It was also crucial to the rationale for constructing an Alaska pipeline destined to bring gas to the rest of the United States at the equivalent of $50 to $60 per barrel of oil. And it has been the major rationale for building synthetic gas plants such as the Great Plains project.

However, as the gas industry is learning, natural gas, in general, cannot compete with oil as a transportation fuel because it requires either liquefaction and cryogenic storage equipment or pressurization and storage capacity roughly twice the size of a gasoline or diesel tank for the same energy content. Therefore, by virtue of its energy density, oil has a virtual monopoly on the transportation market. That could change if someone were to develop an easy way of using residential natural gas to fill up automobile tanks. But even then, that would probably put downward pressure on gasoline prices, not allowing natural gas prices to rise.

Furthermore, as long as crude oil prices are at about their current level or higher, fuel oil made from that crude oil will not be competitive in buildings markets. To hold market share, gas must therefore compete with electricity (and the electric heat pump) in the residential and commercial buildings markets, ensuring that its market price will be a function of the price of electricity, not oil. In industry, gas must compete with coal and residual oil (the oil left over after refining crude oil). But coal prices are kept low because there is so much supply, making it necessary for both gas and residual oil to compete with coal. Because of this competition, residual oil is set at whatever price coal and gas will allow, not the reverse. And if the price of residual oil gets too low relative to market prices of other refined oil products (gasoline and distillate), refiners will make more of the premium product from the residual oil through coking or more advanced upgrading processes. Unlike the transportation market, the competition among fuels is fierce in both buildings and industries.

If oil prices should rise substantially in the future (for example, because of a resurgence of oil demand or a supply disruption), it is unlikely that natural gas prices would follow in lockstep. Higher oil prices would reduce oil use in buildings and industries, but natural gas must then compete with coal and electricity, not oil, to maintain market share. If natural gas were priced equal to oil under these conditions, a gas "glut" might develop—gas would steadily lose market share until the quantity sold was well below available supplies. To maintain their sales at desired volumes, producers and pipelines would lower gas prices to levels where gas-based energy services were competitive with services from these other fuels. In such a market, the market-clearing price for natural gas would be well below oil-equivalent prices. The abundance of competition in end-use markets makes the simplistic assumption that natural gas competes only with oil (and therefore will be priced in line with oil products in the future) a myth.

CONCLUSION

These eight beliefs constitute what we think are the great energy myths of the post-1960 period. Each of them has been the basis not only for ineffective political responses but also for the public's inability to understand the energy problem. Taken together, these myths may explain much of the public's early dissatisfaction with national energy policy. The pronouncements lacked the basis for widespread acceptance. Fortunately, discarding these beliefs is becoming much more commonplace, and progress toward a solution is now more evident and predictable. The attention being given to energy services instead of energy units is shifting the nation's focus from shortage to competition—and abundance.

9

The World View

In previous chapters we have focused exclusively on the positive impacts individual energy decisions are having on the U.S. energy situation. But the United States represents only about one-third of world energy use and a similar fraction of the world economic system. How will future projections of U.S. energy supply and demand affect the world energy situation? Are similar results occurring outside the United States? Will U.S. progress have a positive impact on world energy markets?

WORLD OIL MARKETS

The United States is linked to the rest of the world economically, strategically, and politically through trade, and one of its largest traded items is oil. The United States imported almost $80 billion worth of oil in 1980. Therefore, about 3 percent of the U.S. GNP that year went abroad to pay for foreign oil. Two-thirds of that, or over $50 billion, went to OPEC.

Table 9.1 shows world oil trade in 1980, and three hypothetical projections for the year 2000. About 50 million barrels per day (MMBD) of oil was produced and consumed in 1980 by the non-Communist countries. The bulk of the oil (over 75 percent) was consumed in the OECD countries, which represent most of the developed nations.[1] And

[1] The members of the Organization for Economic Cooperation and Development (OECD) are Australia, Austria, Belgium, Canada, Denmark, Finland, France, the Federal Republic of Germany, Greece, Iceland, Ireland, Italy, Japan, Luxembourg, the Netherlands, New Zealand, Norway, Portugal, Spain, Sweden, Switzerland, Turkey, the United Kingdom and the United States.

TABLE 9.1 World Oil Supply and Demand (million barrels per day)

	1980	1982	2000 Low demand	2000 Medium demand	2000 High demand
World GDP[1] growth (%/year; 1980-year)	—	—	1.7	3.0	3.9
World oil price (1982 $/bbl)	40	33	30	53	80
Oil consumption (MMBD)					
OECD	38.5	34.6	34.5	31.3	28.1
U.S.	(17.1)	(15.3)	(13.6)	(13.5)	(13.5)
Non-U.S.	(21.4)	(19.3)	(20.9)	(17.8)	(14.6)
OPEC	2.7	2.9	4.7	6.5	8.4
LDC[2] (Non-OPEC)	8.3	8.2	10.1	14.0	16.8
WOCA[3] subtotal	49.5	45.7	49.3	51.8	53.3
Non-OPEC production					
OECD	15.7	15.9	12.4	14.2	16.4
U.S.	(10.8)	(10.7)	(8.3)	(9.5)	(11.0)
Non-U.S.	(4.9)	(5.2)	(4.1)	(4.7)	(5.4)
LDC (Non-OPEC)	5.8	7.0	9.0	9.3	9.6
Net CPE[4] exports	1.2	1.5	—	—	—
Non-OPEC production	22.7	24.4	21.4	23.5	26.0
Inventory drawdown	−1.0	1.5	—	—	—
Balance					
OPEC production[5]	27.8	19.8	27.9	28.3	27.3

[1] Gross domestic product.
[2] Less developed countries.
[3] World outside Communist areas.
[4] Centrally planned economies.
[5] Projections show implied OPEC production at different levels of demand and prices. Desired OPEC production is estimated at 28 MMBD in 2000.

almost 60 percent of the oil consumed in the OECD was imported. The less-developed countries were also net importers, consuming about 50 percent more than they produced. The OPEC countries consumed only a small fraction (10 percent) of the oil they produced in 1980.

On the production side, the balance is much different. In 1980, OECD countries produced almost one-third of the world production (15 MMBD), the majority of which was produced in the United States. Non-OPEC less-developed countries (LDCs) produced about 6 million barrels per day. And the balance of 28 MMBD was provided by OPEC—well over half of all the oil consumed in the free world.

The balance between oil producers and consumers can shift dramatically and rapidly, as the events since 1980 have demonstrated. For example, Table 9.1 compares 1980 oil trade with 1982. Since 1980 total free-world oil consumption has declined about 8 percent, owing to the rapid price increases in 1979 and 1980 and the subsequent worldwide recession. Almost every nation has reduced its oil consumption since 1980. Higher prices have also stimulated an increase in non-OPEC oil production by over 1 million barrels per day (7 percent), and excess supplies have caused consuming countries to draw down their inventories. All of these factors have worked to squeeze OPEC production out of world oil markets. Because OPEC acts as the marginal supplier of oil, the demand for OPEC oil (measured in Table 9.1 by OPEC production) has dropped 30 percent since 1980 as the world oil market adjusted.

As in all commodity markets, the major adjustment mechanism for the world oil market is prices. Although OPEC obviously has some control over prices, the world oil market has recently behaved surprisingly like other commodity markets, where the dynamics of supply and demand are the major determinants of prices. Therefore, to project oil prices, we estimated how the world oil market might balance at different hypothetical future demand levels. Because future world economic growth is so uncertain—and oil demand will likely depend on how fast world economies develop—we hypothesized three scenarios of world GDP growth to the year 2000 (1.7, 3.0, and 3.9 percent per year, as shown in Table 9.1).

In the medium demand case, world economic growth is expected to average 3.0 percent per year from 1980 to 2000. Using our Least-Cost projections for U.S. oil demand, and extrapolating the U.S. results to the rest of the OECD, we concluded that OECD oil demand would most likely remain below 1982 levels through the end of the century. OPEC and LDC oil consumption, however, is expected to increase over this period, as their economies develop. Historically, oil demand for this group has grown at 7 percent per year, even during the period 1973–1980, when oil demand in the developed world stagnated. Depending on LDC economic recovery, we project non-OPCD oil demand to grow at 2 to 5 percent per year over the 1982–2000 period. Total World Outside Communist Areas (WOCA) consumption might therefore range from 49 to 53 MMBD in 2000, depending on world economic growth (for comparison, 1980 consumption was 49.5 MMBD).

On the supply side, U.S. oil production could drop about 25 percent below 1982 production levels by the year 2000 or could remain roughly constant, depending primarily on the projected level of oil prices.[2] Total

---

[2] Estimates of world oil production were derived from Department of Energy sources: *The National Energy Policy Plan* (U.S. Department of Energy, *Energy Projections to the Year 2000*, July 1981) was used to derive projections for U.S. oil production, and production figures for the rest of the world were obtained from the Energy Information Administration, *1981 Annual Report to Congress*, vol. 3, February 1982.

OECD production (70 percent of which is U.S. production) ranges from 12 to 16 million barrels per day in the year 2000 (compared to 15.9 MMBD in 1982), depending on price. LDC oil production includes Mexico and will increase significantly as these countries use their oil resources to develop their economies, even if world oil prices remain relatively low in the future. Exports from centrally planned economies (CPEs)—primarily the USSR—will probably not add to free world supplies in the future, so the total non-OPEC oil production will range from 21 to 26 MMBD over the next 2 decades under the three demand scenarios.

To project a likely path for world oil prices, we assumed, as others have, that OPEC would want to maintain production at a desirable, or "equilibrium," level of capacity utilization (where production is roughly 80 percent of capacity). We then projected OPEC's production potential from Department of Energy estimates of future OPEC capacity and obtained a result that OPEC might want to increase production back to 1980 levels over the next 2 decades—to about 25 to 30 MMBD. In the medium demand case, in order for OPEC to produce at desired levels (28 MMBD), world oil prices will need to rise to no more than about $50 per barrel (in 1982 dollars) in the year 2000. Depending on world economic growth, prices might range from $30 to $80 per barrel in that year. For comparison, world oil prices were as high as $40 per barrel in 1980 (in 1982 dollars) and have dropped to $29 per barrel in 1983.

These projections, illustrated in Figure 9.1, suggest that the price rise of 1979 was significant enough to create a slack oil market for roughly a decade (1980–1990). Over the longer term, however, the recovery and sustained growth of world economies should add enough to world demand to cause oil prices to rise above 1980 levels. In the medium case, we estimate that oil prices will rise at an average rate of 1.5 percent per year to 2000, the major portion of which will occur in the 1990s.

All projections of the world oil market, including these, are extremely tenuous, for oil prices are relatively sensitive to even slight changes in supply and demand assumptions. For example, U.S. and other OECD oil production could vary from our projected levels depending on the success or failure of future drilling. OPEC production will depend on OPEC's revenue needs, production constraints, political considerations, and even the condition of world financial markets. Oil production in less-developed countries might vary depending on geological, economic, and political factors that are hard to predict. Total oil supply might easily be 5 million barrels per day higher or lower than we project, depending on factors other than oil prices. Similarly, on the demand side, economic growth, production of other fuels, and especially the consumption of oil in less-developed countries could shift demand by

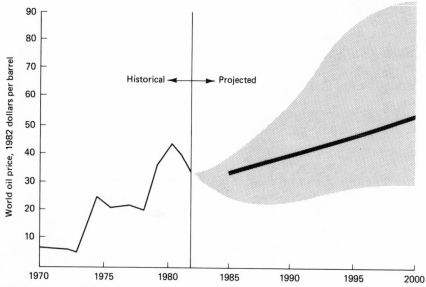

Figure 9.1 World oil prices.

another 5 MMBD. These uncertainties could cause larger swings in future oil prices, as illustrated by the shaded area in Figure 9.1. If supply or demand shifted by a total of 10 MMBD in the year 2000, projected year 2000 world oil prices might vary from $30 to $80 per barrel.

Furthermore, these projections represent equilibrium oil prices—and it is more likely that the world oil market will be out of equilibrium more often than not, as has happened over the past decade. Although a relatively smooth time path for world oil prices is shown in Figure 9.1, the most likely path for oil prices is probably rapid price increases in tight-market periods, followed by stable or declining prices in the slack market that inevitably follows. But because OPEC has been badly burned by their excessive price increases in 1979, such swings should be much less pronounced than in the past. Despite these potential swings in price, we continue to be persuaded that the basic competitive forces affecting the world oil market will imply the lower side rather than the higher side of these price projections.

OIL VULNERABILITY

Compared to other oil-importing countries, U.S. dependency on imported oil is relatively modest. Whereas imported oil reached a maximum of 22 percent of total U.S. energy consumption in 1978, Japanese dependency peaked at three times that of the United States, a whopping 75 percent in 1977. Similarly, West Germany was 57 percent dependent in 1980, and the whole of Western Europe was 62 percent dependent in 1973.[3] Of the OECD countries, only the United Kingdom, Canada, the Netherlands, and Norway have a dependency figure lower than the United States. Thus, regardless of how successfully the United States reduces its import dependency in the future, the majority of our Western allies will need imported oil for some time. But the United States, being the largest single consumer, can increase or decrease the vulnerability of OECD countries to oil disruptions depending on its oil consumption and production.

The vulnerability of non-OPEC developing countries is also greatly affected by U.S. requirements. These troubled countries account for about 15 percent of all energy consumed in the free world, less than half that of the United States and one-fifth of the OECD countries. In spite of this disparity, the major difference between developed and developing countries is that the opportunities for reducing consumption in developing countries is small, since they lack the capital to bring about more than a minor change in energy use. And consequently, energy costs in these countries are largely dependent on the actions of others. If the Western industrial nations can turn off the oil spigot through energy efficiency improvements and alternative-energy supplies, LDC nations will be major beneficiaries.

Fortunately, all the industrialized nations are responding to higher oil prices with major efficiency improvements. U.S. consumption of oil per dollar of GNP dropped 16 percent from 1973 to 1980, while the average for all OECD countries dropped 20 percent. Since other industrial countries appear to be experiencing the same energy productivity phenomenon seen in the United States (see Table 9.2), the energy outlook for the developing countries is more promising.

TABLE 9.2 Index of Energy Consumption (Ratio to real GNP)*

	United States	Japan	France	West Germany	Italy	Britain	Canada
1973	100	100	100	100	100	100	100
1980	86	82	90	85	88	88	103

* This index means that if it took 100 barrels of oil equivalent to produce a given amount of GNP in 1973, that same amount was, for instance, produced in the United States by 86 barrels in 1980.
SOURCE: International Monetary Fund, "World Economic Outlook," June 1981, p. 147.

[3] Daniel Yergin and Martin Hillenbrand, *Global Insecurity: A Strategy for Energy and Economic Renewal,* Houghton Mifflin, Boston, 1982, p. 202.

The Effect of Oil Prices on the World Economy

To be sure, the economic losses resulting from the oil price hikes in 1973–1974 and again in 1979 have been awesome. In the seven largest industrial economies, for example, cumulative inflation was 112 percent between 1974 and 1981 versus only 44 percent from 1966 to 1974, and the cumulative economic growth was only 21 percent versus 46 percent in the earlier years.[4] The OECD's chief economist estimates that GNP losses from the 1979 interruption and price increase were $360 billion in 1980 and $620 billion in 1981, roughly 5 to 8 percent of the free world's GNP.[5] But the real question relates to the probability or even possibility of similar disruptions in the future. And the opinions range from absolute certainty that there will be another major disruption to absolute certainty that there will not.

The Probability of a Future Disruption

Witness two contrasting views: one from Frank Potter, staff director of the House Committee on Energy and Commerce, and the other from Fred Singer, Senior Fellow at the Heritage Foundation (while on leave from the University of Virginia). Potter says: "Do we really have to worry about [another] energy crisis? Imports are down and OPEC seems to be in trouble—does this mean that our recent energy policies have been successful? Comforting as this simple reasoning may be, it is also *wrong*."[6] On the other hand, Singer says:

> Many people do not realize that we have become, in effect, *immune* to oil embargoes. And so, they continue to worry about our relations with oil producers. In my opinion, the Saudis could do us a great favor by announcing an embargo, and thus demonstrate once and for all that these worries are groundless. But my advice would be not to try it—it would kill their potential clout and cause a large loss of revenue. They cannot afford either.[7]

[4] Daniel Yergin and Martin Hillenbrand (eds.), *Global Insecurity: A Strategy for Energy and Economic Renewal*, Houghton Mifflin, Boston, 1982, p. 6.

[5] Sylvia Ostry, "A View from the OECD World," paper presented at the 20th Anniversary Meeting of the Atlantic Alliance, Oct. 20, 1981.

[6] Frank M. Potter, Jr., paper presented to the American Bar Association, Aug. 9, 1982. Others agree with Potter: for example, Daniel Yergin and Martin Hillenbrand, op. cit.; and Charles K. Ebinger, *The Critical Link: Energy and National Security in the 1980's*, Ballinger, Cambridge, 1982.

[7] S. Fred Singer, "Oil Supplies Depend on the Market, Not on Politics." *The Wall Street Journal*, Aug. 27, 1982.

We believe the truth is somewhere in between. Singer is right that the market has diminished the threat of an embargo. But he is terribly premature in announcing embargo immunity. Potter is right that our vulnerability remains immense; the 1973 and 1979 price increases have shown that the price of oil is volatile and that small shortfalls can translate into large price increases. History shows that the price of oil almost tripled between 1979 and 1981 from an actual shortfall of less than 5 percent. But Potter and others need not go so far as to announce a third oil shock to make their point—that we need proper insurance even for a "remote" possibility.

As we have indicated before, we do need that insurance, both for our own economy and for that of our allies. Even if the overpricing of foreign oil reduces U.S. oil imports to low levels in the future, we remain vulnerable through our interdependence with the rest of the world. Even if we are not negotiating large quantities of oil in the future, our trade partners will be—and an oil cutoff would affect us through them. Our interdependence calls for our assistance in the event of a discontinuance of a portion of the world oil supply. As a member of the International Energy Agency, we would indeed be obligated to cut consumption to provide that assistance in a shortfall exceeding 7 percent.[8] But as Alvin Alm says: "The lack of an agreed-upon measure for pricing transactions under the agreement raises serious questions about the feasibility of the IEP [International Energy Program] system," even between the 7 and 10 percent shortfall range.[9] Above 10 percent the agreement's inequities could lead to bitterness and ultimately to amendment or dissolution of the sharing agreement.

This all leads Potter to assert that "at the very least, it would be useful to develop policies at home and with our allies which would lessen the need for emergency legislation." He further suggests that the Strategic Petroleum Reserve continue to be built up from its 1983 level of about 300 million barrels to triple that amount and that precise rules be worked out as to when the reserve can be used.

We and almost all others we know concur with these suggestions and it would be foolish not to. From that point, however, the agreement dissipates. Potter suggests that we must also have an emergency plan that goes well beyond the "free market." "The President should reconsider the use of limited price and allocation controls in an emergency."[10]

---

[8] The IEA sharing formula is based upon members' relative oil consumption for a shortfall of between 7 and 10 percent. Thereafter it is weighted toward the relative amount of imports.

[9] Alvin L. Alm, "Energy and the Western Alliance," Aspen Institute for Humanistic Studies, 1982, p. 12.

[10] Frank M. Potter, Jr., op. cit.

Others, ourselves included, point out that such programs did not work at all in the 1970s and there is absolutely no agreement as to a series of such measures that would work in the future. Even Potter states that "the main problem with these approaches has been experienced in their implementation."

We, would therefore, urge that the market be the allocator of short-term oil shortages—if they do occur—as well as long-term supply and demand relationships and that any ideas of allocating or controlling by regulation be abandoned. *All* of our allies used the market in the embargo of 1973 and during the Iranian revolution in 1979 and fared better than the United States. Odd-even purchasing plans, rationing, and price and allocation controls have all proven cumbersome, ineffective, and very likely more inequitable than allowing the market to sort things out. With a solid strategic petroleum reserve to dampen major price swings and a large overhang in world production capacity, a market strategy would be much less drastic than massive market intervention.

An unexpected threat to the West is the new possibility of a significant reduction in OPEC oil prices. For instance, we tested in our models an assumption that oil prices remained at $33 per barrel (1982 dollars) instead of rising to over $50 per barrel in 2000 as in our Least-Cost projections. In that case we found that U.S. import dependency almost doubled compared to our Least-Cost case. This means that with lower prices, the Strategic Petroleum Reserves would have to be increased.

Of course, the attendant benefits of lower oil prices are attractive. For instance, the American Express International Banking Corporation estimates that Brazil would gain some $2.5 billion in their import payments from an oil-price drop to $25 per barrel.[11] Economic growth would be slightly better (the OECD says a $25 per barrel price would increase GNP growth from 1.5 percent to 2.25 percent in 1983 for its 24 member nations), energy-intensive industries would be more profitable, transportation industries would increase sales, and inflation would be lower (OECD says it would drop by 1 percent). But this is exactly what led to the embargo in 1973. Low oil prices—or more accurately, lower-than-market oil prices—stimulated oil demand to the point that all oil-using countries became vulnerable to oil imports.

As we have seen, a high-oil-price future reduces free-world vulnerability. If prices are below market for long, the United States and its allies will again become vulnerable to oil import disruptions. So that history doesn't repeat itself we would advocate a U.S. oil import tariff that is

---

[11] American Express also estimates Venezuela, Mexico, and Nigeria would lose $2.5 billion of revenues from a $25 per barrel price in 1983.

triggered by a vulnerability measure—perhaps the absolute volume of oil imports relative to total U.S. oil consumption (see Chapter 10 for an expansion of this idea).

A circumstance where such a policy would have been extremely useful occurred in early 1983. The OPEC ministers were unable to reach agreement about production quotas or prices. The smaller producers wanted to maintain a price advantage on the larger producers, mainly Saudi Arabia. The threat was that if there was no agreement, the Saudis would cut their price and OPEC would be in chaos. If the United States had a policy of taxing imports above a certain level, the effects of such a price cut would be significantly muted. As it happened, OPEC did cut its price $5 per barrel, in an attempt to stimulate world oil consumption and dependency. Undoubtedly, as a result, U.S. and other OECD imports will go up. Without an import tariff policy, other price cuts may be instituted, and soon we could be vulnerable again. Despite a drop in oil prices after 1980, it is difficult to foresee a circumstance in which the United States and its oil-consuming allies return to pre-1973 levels of oil dependence and vulnerability. The response on the demand side has been too strong and is too permanent for the 1973 ways of using and wasting energy to return. Even a recent Harris Poll showed that U.S. consumers did not intend to change their plans because of OPEC's current weakness.[12] This is likely the case in other consuming countries. Like the United States, the world is generally accommodating the energy transition very well indeed.

[12] "Energy Guzzling: Most Consumers Are Cured," *Business Week,* Apr. 4, 1983, p. 16.

10

What Else Is Required?

The capacity to achieve energy abundance is within our grasp. It is taking the form of a private, varied, and individual response to the opportunities . . . It is involving the participation of manufacturers, contractors, state utility commissions, bankers, building code jurisdictions, utilities . . . all consumers, and government. The only thing missing so far is our confidence that it is happening and our commitment to further the process.

The previous chapters have shown that the nation, and probably the world, is moving more rapidly than anyone expected through its most difficult energy transition—propelled by a dizzying variety of energy productivity techniques. Innovative methods and systems proliferate across the land, reducing the cost of energy services to the consumer. What else, if anything, is required before there is sufficient assurance that the transition will be completed in a reasonably timely manner and in a way that will maximize the benefits to the American people? What ingredients are missing from the formula that is shifting the flow of energy investments by individuals and companies away from conventional supply patterns into products and services that minimize the costs of energy services? In short, are there any other actions, in this era of high fuel prices, that are required to ensure that consumers participate fully in the new cornucopia of lower-cost energy services?

GOVERNMENT SHOULD REDUCE ITS ROLE

Until recently the traditional response to these questions would have been yes and the requirement was for some form of government intervention. Historically, people have looked to government, especially at the federal level, for programs and actions to protect them during most major economic and social upheavals. Energy has been no exception. Yet while government institutions do have a role, there is little evidence in the evolution of U.S. energy policy that suggests a federal role produces the greatest good for the greatest number. One of us vividly remembers the hours soon after the Arab oil embargo when Nixon administration officials feverishly searched for ways to demonstrate to the nation that the president had the situation under control. The leadership decided to create a massive energy development program comparable to both the Manhattan Project (which produced the atomic bomb) and the Apollo Project (landing men on the moon). In those frantic days a contest was held to select a name for the program. Herb Stein of the Council of Economic Advisors came up with Project Independence as the winning entry. The entire Project Independence program was developed in less than 2 weeks so the president could announce it to the nation from the White House. The goal he announced was "self-sufficiency in energy by 1980."

This blurred snapshot of frantic energy policy development underscores how little the experts knew about the long-range energy situation facing the nation. Even as the president announced his goal, only a handful of his staff realized that petroleum self-sufficiency in 7 years was absurdly unrealistic. Furthermore, the detrimental side effects for the economy and the environment, even if that goal could have been achieved, were consistently underestimated. The policy makers' entire attention centered on the embargo's effect on the economy and on ways to quickly reduce the high level of oil imports. This particular focus quickly led to the formation of some basic premises (itemized as myths in Chapter 8) upon which the bulk of all energy policy was formed during the next 7 years, 1973 to 1980.

Nevertheless, most of the controversy and difficulties regarding the government's past role in energy have not revolved around the conflicts between energy goals and other public goals or even in the setting of energy goals themselves. Rather, they have been with the means of implementation. Most of the energy goals formulated during the past 10 years were admirably designed for specific purposes: minimizing a particular environmental hazard, reducing vulnerable supplies of oil, lowering the price of fuel to the poor, or preparing for the next embargo. Those that argue against such goals have a perception of the world that is difficult to comprehend. On the other hand, past energy policies sought to achieve those goals primarily through regulation of one part or another of a complex system. And these regulatory approaches were almost always based on false premises. Because the lawmakers and regulators' knowledge of the energy system and market was (and is) so limited, they usually omitted or even prohibited some of the most attractive energy solutions while encouraging options that proved to be unwise. That approach has not aided the energy transition. As the Energy Information Administration's report said,[1] the positive effects have come primarily from one action; deregulation of oil prices.

All this leads us to suggest that for the future, we first establish an energy policy goal that is consistent with what people are doing: *to ensure the delivery of appropriate quality energy services at the least cost.* Moreover, we believe that the principal means of implementing this Least-Cost goal and balancing this with other social goals should be market forces that motivate people to act in their own interest. In modern politics this implies a unique combination of the most effective platforms of liberals and conservatives: the liberals' social goals implemented by the conservatives' market forces.

Earlier chapters of this book have outlined some of the energy choices available to achieve this goal and the likely energy futures that might result. What specific policy actions would this Least-Cost goal imply? All the policy recommendations that follow imply less, rather than more, government involvement in energy markets.

RESTRUCTURE GAS AND ELECTRIC UTILITIES FOR GREATER COMPETITION

As we suggested in Chapter 8, the current distribution and production functions of utilities should be separated. The criteria for monopoly, namely, the obligation to serve and decreasing costs from larger plants, no longer hold for electric producers (as opposed to distributors), and never did hold for gas producers. Gas and electricity can be produced in so many different ways and by so many different suppliers that regulating the generation of electricity or the production of gas reduces consumer choices, raises costs, decreases flexibility in responding to new conditions and increases government intervention unnecessarily. Policymakers need only note the positive response to the recent shift in federal

[1] U.S. Department of Energy, Energy Information Administration, *Energy Program/ Energy Market*, July 1980.

policy, which allows alternative technologies like cogeneration systems, windmills, and small hydroelectric plants to compete with utility-generated electricity, for a slight indication of what would happen if generation monopolies disappeared.

Federal and state governments should regulate only the distribution of gas and electricity, not its production. The distribution function could be performed either by government or by regulated private enterprise, just as some highways are publicly owned and maintained and some are private (e.g., toll roads). The price charged by these utilities should be a toll, based on the cost of distribution. This regulated toll should have no direct relationship to the value of the commodity being distributed (gas or electricity), just as the owner of a toll road charges no higher fare for a Cadillac traveling the road than a Ford. Such a restructuring would allow federal and state governments to rationalize electric and gas distribution systems, providing more flexibility for wheeling power and distributing gas to where it is needed most.

Just a few years ago, the prospects for such a fundamental policy change appeared remote. There are many difficult political problems involved in the deregulation of utilities, especially the preeminent role of state governments in this area. But the continued slide in the financial health of utilities as presently structured, the emergence of competition from cogeneration and conservation technologies, and the efforts by many utilities to restructure themselves along energy service lines by creating unregulated subsidiaries all argue for a fundamental change in regulatory policy.

REPEAL RESTRICTIONS AND DISINCENTIVES FOR FUEL USE

Logically, if the nation enters an era in which choices for providing energy services abound, the government need not establish rules regarding which fuels should or should not be used for particular purposes. The federal Fuel Use Act of 1978 is especially troublesome in this respect. Such restrictions are necessary only when price controls lead customers to make uneconomic choices about fuel. Laws that require utilities to switch from oil and gas to coal are also unnecessary, especially if the utility producing electricity were unregulated and forced to compete in the marketplace.

Laws requiring energy-consuming or energy-conserving technologies to meet certain environmental and safety standards are perfectly appropriate, especially if the standard is maintained by a fee or other economic incentives. Some of these requirements will be less costly with one technology than with another, but the choices are becoming so plentiful that we need not be concerned that current environmental laws will hinder delivery of future energy services at reasonable costs.

IMPOSE TAX ON EXCESSIVE OIL IMPORTS TO PAY NATIONAL SECURITY COSTS

Imports, even oil imports, are obviously not in themselves bad for the country. In fact, if oil or any other necessary product can be purchased from another country cheaper than it can be produced in the United States, it is better to buy it from outside. The problem arises when the quantity of imports reaches such a level that arbitrary supply disruptions can cause economic hardship, even if the hardship is short-term, while the economy adjusts to the disruption by adopting substitutes for those imports. When imports reach that level, say, more than 15 percent of oil consumption, national security and economic costs are incurred.

As argued in Chapter 9, we think that an import tax is called for. The tax would depend on the quantity of imports, not the price, but only after imports reached a point where a national security cost is incurred that would increase as the import share grows higher. The amounts collected could pay for increasing the Strategic Petroleum Reserve and defense spending (i.e., the Middle East fleets) required because of excessive dependency on imported oil. If imports were below 15 percent of oil consumption, no revenues would be raised. If imports were high, the revenue raised would be commensurate with the national security cost. Finally, this government action would place the cost of meeting the nation's national security goals on the consumers of the product that caused the security risk in the first place.

ELIMINATE SUBSIDIES AND TAX INCENTIVES

Consistent with its goals, the government should repeal those laws that give tax or other benefits to any one particular fuel or conservation technology. Although these subsidies are difficult to distinguish from general business or individual tax incentives, some are obvious. Tax incentives for solar energy systems, gasohol and insulation certainly fall into this category. Nuclear industry insurance, government uranium enrichment facilities and waste handling that is paid for by the government, instead of by the consumer of the electricity generated at these plants, is a blatant subsidy that skews the competition among choices available to the consumer. The same is true for production loan guarantees and other subsidies provided for synthetic fuels by the Synthetic Fuels Corporation.

The law that gives municipal utilities first call on low-priced hydropower is another such subsidy. In fact, all the federally owned hydropower that is sold to customers at prices far below its value (e.g., in the Pacific northwest) is a subsidy that causes consumers to choose high levels of electrical use instead of those conservation and other production technologies that could lead to a much better use of resources.

These subsidies are all exceedingly difficult to repeal. Powerful interest groups have grown dependent on the subsidies in the law. But the role of government should be to ensure that the options available to deliver energy services are given an equal and fair chance to compete. Subsidies tend to blur the choices and often make one choice appear attractive to a consumer (i.e., federal hydroelectricity) when, in fact, the real cost or value of that choice, as indicated by the cost of new nuclear or coal plants, without the subsidy would diminish its attractiveness. The subsidy issue again puts the government into the position of making choices about technologies instead of letting intelligent, informed consumers make up their own minds. Furthermore, there is little evidence to suggest that societal goals have been better met as a result of such subsidies, and there is no evidence in our forecasts (in Chapter 7) of any need for subsidies in the future.

REORIENT GOVERNMENT EXPENDITURES ON ENERGY RESEARCH TOWARD PERSONAL CHOICE

Government spending for *energy research* can be justified to ensure that the American people have as many energy choices in the future as possible. Yet the technologies that receive government funds should not be selected by government bureaucrats. Choosing potential technology "winners" early in the research stage is difficult, but there are few people less qualified to make the choice than the legislative and executive branches of government. As late as 1980, approximately 80 percent of all Department of Energy research and development funding was being spent on large-scale central-station electric-generating technology when there were strong indications that during the next 20 years large power plants would be among the least-competitive energy service options— second only to imported oil. If the extraordinary bias toward big electric technologies were eliminated, we would support continuing fundamental *research* in nuclear or other electrical-generating technologies along with other pure-energy research. But federal development of these technologies into commercial operations is a trap that can be no better illustrated than by our sad experience with nuclear power.

CREATE AN ENERGY EMERGENCY PLAN

One of the primary concerns of government policy has been planning for another oil embargo or similar crisis. Yet most types of measures to temporarily curtail energy consumption in an emergency that have been considered by and proposed to Congress have no more chance of working than the allocation systems used in 1973.

Clearly, preparing the country for emergencies is a legitimate government role. Even if government bureaucrats are inadequate substitutes for marketplace decision makers, they will be forced into action by a public requiring its government's involvement in a time of crisis. That was the case in the fall of 1973 when the Arab embargo brought substantial public pressure on the Nixon administration to respond.

The problem is how to prepare for such emergencies. The Department of Defense spends billions of dollars preparing for emergencies and setting up contingency plans for every significant national security threat, however remote. Even though the potential of an energy crisis is substantial, we conclude that no such elaborate plan is justified. A simple use of strategic petroleum reserves to dampen severe market reaction should be adequate if supported by good government data on energy use and production (such as the Department of Commerce provides on the economy).

These six areas of government action are not the extent of all that government should do in energy. They represent only those areas where change appears most appropriate. Other ongoing government efforts like prudent leasing of coal and oil fields on federal lands are consistent with the goal of making maximum choice available as outlined earlier.

What is needed is acceptance of a government role that does not decide which fuels or energy sources should be produced and which should be saved. Nor should government policy assume that there is any inherent value in producing or saving energy—only in satisfying the energy services needs of the country.

In summary, the government's role in the future should be to ensure a competitive environment that provides energy services at the lowest cost. The strategy should be: (1) to make the maximum number of choices available in all energy service markets; (2) to ensure that markets are unhindered in providing proper signals about the costs and benefits of those choices so that these alternatives for meeting the needs can compete fairly; and (3) to use fees, taxes, or incentives when free market results are inconsistent with other public goals such as minimizing oil supply vulnerability or environmental damage.

PRIVATE SECTOR COMPANIES SHOULD EXPAND THEIR ROLE

As the previous section demonstrates, the real opportunity today is not susceptible to government solution. It is being realized by expanding the flexibility and choices available to consumers in obtaining energy services. To further broaden the options will require more new business activity specifically oriented to providing energy services at a competitive cost—energy service companies. In spite of all the progress that has been made, the opportunities for such companies remain enormous. As we have seen, making space comfortable in buildings is still usually predetermined by a building developer, not by an expert on energy services seeking to minimize the annual cost of such comfort. Consumers must still buy natural gas and electricity from utilities that have a monopoly on their commodity and that determine price solely on the basis of the utility's cost of distributing gas or electricity. For the most part, residential customers pay the same for each unit of fuel whether they use 10 times more or less than the amount used by their neighbor. Cost-related quantity discounts do not exist. Most consumers still pay the same whether they use all their electricity or gas at night or in the daytime. Even summer and winter use does not alter their costs in most places. And, they cannot negotiate a better deal if they spread their demand evenly throughout the year versus purchasing most of it during a particular season (e.g., for a vacation home).

Finally, only with great skill, substantial research, access to financing, and the ability to secure and manage contractors can consumers ensure that their houses or businesses have the most economical system to convert the gas and electricity they purchase into needed energy services.

Why are the choices more limited than they need be? Because historically natural gas companies have marketed gas, not units of heat or comfortable rooms. Similarly, electric utilities have sold electricity, not light, heat, or convenience (even though light was the original product). And because of close regulation and a traditional emphasis on engineering, marketing approaches that would provide choices to serve customer needs more precisely have not been forthcoming or have been missing for years.

Residential customers are not alone. In industry, the traditional energy companies have done their marketing in much the same way. With some exceptions, the typical industrial consumer is tied to procuring fuels and electricity from utilities. In order to reduce the cost of converting the energy into useful services like heating metal, molding tires, or drying paint, customers first must develop expertise on possible alternatives. Then they must secure financing for any new equipment required and manage the new equipment as part of their operations. For some businesses this is now a way of life. Others, however, seem to be searching for a different service. "That's not our business," said a president of a chemical company when advised of an opportunity to reduce the cost of energy services by making a large energy productivity investment in his plant. "If we have that kind of capital available to us, we are going to put it into new chemical products."

In transportation, the scenario is identical. Personal mobility is obtained by purchasing a car—typically one that optimally fits a person's most important mobility needs. Next there is bank financing, gasoline purchases from an oil company's retail outlet, maintenance from a garage, insurance from an insurance company, and licenses and permits from a government. No one as yet, other than short-term rental companies, offers a one-step service to minimize the cost of all these components. But it will be done. It is waiting to happen.

To be sure, consumers have had considerably more choice in the personal-transportation area than in some of the other services, since there is a large selection of vehicles and gasoline stations to choose from. Auto selection is also something with which many consumers feel very comfortable. However, higher gasoline prices and higher prices for new high-mileage cars are changing the economics of the services in this sector and developing new ways to market mobility—on a cost per mile basis—are needed.

As we have shown repeatedly, the market for new energy service businesses has not gone unnoticed. Energy service companies are being created all over the United States. Industrial concerns that lack the technical skill, the financing, or the inclination to be in the energy business are being offered electricity and steam produced in a private, unregulated cogeneration system to reduce the cost of those commodities. Typically, the company offering to sell the cogenerated steam purchases, installs and operates the equipment. But this is not a new idea. For example, in 1962, Air Products, Inc., recognized such a market opportunity. For years, Air Products supplied industrial customers with oxygen, nitrogen, and other gases that they manufactured at a central facility and then shipped to customers. A number of years ago they changed strategies and began to build some of their manufacturing facilities on the site of their industrial customers. Air Products was able to offer the customer lower prices for the manufactured gases because transportation costs were eliminated. The industrial customer got the lower prices without having to become an expert in a new business or shift capital away from its primary business. Many entrepreneurs are studying that experience today with regard to energy.

Likewise, as energy service costs have risen and the task of reducing these costs has required too much capital and greater technical expertise, owners of large office buildings, schools, and hospitals are exploring the option of transferring the task of heating, lighting, and cooling of those buildings over to companies that specialize in delivering those services. The new energy service companies contract with the building owner to pay the utility bills and provide the building with specified comfort and convenience services (e.g., 72°F and electricity for typewriters and computers) during the time the building is occupied. This is analogous to the cleaning and food services that most building owners once performed themselves but now contract to firms that specialize in providing those services.

For the nation's residential consumers there exists a void. It can be filled by companies whose wares do not stop with fuel oil, gas, or electricity. It is just too complex and costly for the average home owner to choose from all these alternatives and use them in a way that will reduce his or her overall cost to a minimum. What we are entering is an era in which there is a need for businesses to sell comfortable rooms. Then the seller of the comfort will be responsible for ensuring that furnaces are properly tuned, or that the house is optimally sealed or insulated, or that a water heater is installed that will produce the hot water at the lowest cost.[2]

If such an environment were created, home owners and realtors wouldn't feel that they had to become energy experts to save energy service dollars. As the American Public Power Association says: "Consumers could care less about electricity per se; what they are interested in is being warm in the winter and cool in the summer, having light by which to read and work, and access to motor power to run their appliances and equipment. Customers want to know how they can secure those energy services in an adequate fashion at the least cost."[3]

In some respects, the concept of energy service businesses is similar to the emergence of the concept in the health field of health maintenance organizations (HMOs). As the cost of doctors, sophisticated equipment,

---

[2] We first wrote about this in "Coming Markets for Energy Services," *Harvard Business Review,* May–June 1980, pp. 6–24.

[3] Larry Hobart, op. cit.

and hospitals increased, one way to reduce the cost of health care and keep people healthy was to ensure that the health service received matched the actual health problem. This could be accomplished, for example, by not using a highly trained, experienced doctor to perform duties that someone less expensive could accomplish or by not leaving a person in a hospital room longer than necessary. The conventional approach requires patients to purchase services to make them well after they become ill or, more frequently to purchase insurance that pays doctors and hospital services after the illness had arisen. Under the HMO alternative, individuals contract with HMOs for all health services. The theory is that the HMO has the incentive to keep people healthy or to keep people from requiring expensive treatment. Thus, through preventive care and the efficient use of doctors, paraprofessionals, hospitals, and equipment, an HMO might reduce the cost of health care to the consumer while increasing its profits at the same time.

The case for creating new energy service companies, similar to HMOs in the health field, is applicable to energy consumers in the home, office buildings, and industry. Great potential exists even in transportation. The cost of personal mobility shot up rapidly in recent years because the cost of gasoline increased so dramatically, but other mobility costs have risen as well. As discussed in Chapter 3, one way to reduce the cost of personal travel is to increase the efficiency of the automobile. But this raises the initial cost of the automobile to a point that is prohibitive for many individuals. Over the years as incomes rose and the real cost of purchasing automobiles dropped, it has become popular to own one or two autos that meet all personal transportation needs. The average person purchases a car capable of taking the entire family several thousand miles on vacation, carrying the lumber for a new garage, taking the family out to dinner and transporting him or her to and from work. What is needed, as the cost of ownership rises and maintenance becomes more expensive, are more choices for a family to "buy" transportation mobility for its particular short-term need—a truck for hauling dirt, a station wagon for taking the Little League team to a game, a sports car for a drive to the mountains, and a jeep for the fishing trip.

Most cars available for rental today are of a general-purpose nature, and access is typically only at travel centers like airports and downtown hotels. Will neighborhood companies emerge to market personal mobility on a time-and cost-per-mile basis instead of selling cars, gasoline, and maintenance separately? If so, new options would be open to consumers that would permit them to match more effectively the services offered with their needs and to reduce the total cost of these services.

COMPETITION FOR UTILITIES AND OIL COMPANIES IS ON THE RISE

Our work indicates that the potential for innovative business approaches to providing energy services is immense—the revenues of such companies could total hundreds of billions of dollars over the coming decades. Where are the institutions that will likely take advantage of these energy service opportunities? Will new companies arise or will existing companies adapt? What is the future role of our traditional suppliers of energy such as electric utilities and oil companies?

For their part, most electric utilities in this country are private companies that half a century ago were given exclusive rights in return for an exclusive obligation to serve customers in a certain geographical area. The rationale for this arrangement was that electricity is one of those special services that everyone needs and for which there is not, or should not be, any competition. We have argued that this arrangement is sound for the electric wires that carry electricity to homes and offices; having more than one company digging holes or installing overhead wires is probably unnecessary, duplicative, disruptive, and costly. However, production of electricity is another matter. There are numerous ways to make electricity. Hydroelectric plants, coal, nuclear, oil, and natural gas are the common ones in use today. With increased electricity and other fuel prices, cogeneration is now a competitor, and possibly wind generators and low-cost solar photovoltaic cells will enter the competition over the coming decades.

It is useful to note that in the last few years, the U.S. telephone system has been partially deregulated so that competitors to AT&T can sell phones and other services. Other companies like MCI offer long-distance service in direct competition with AT&T. As a result, numerous choices of communication systems that never existed before are available to customers—enabling customers to lower their cost of service. We believe that the current utility monopolies in the United States are analogous to the telephone system of a decade or so ago. Competition is technically available and should be encouraged. Similarly, excluding the utilities from these new businesses, as some federal and state laws have, is unfair; it is akin to excluding AT&T from the exciting part of the communications market. It will be a painful transition to deregulate power generation after all these years, as it is in the Bell System, but it is just as logical.

Another option to help provide Least-Cost electric energy services is times-of-use rates. Electricity costs less to produce if the equipment producing the power runs continuously near capacity, for this spreads the cost of the equipment over more units of electricity. Hence competitive generators would be willing to provide electricity at lower prices during off-peak periods—just as the telephone company offers cheaper phone rates after 11 P.M. and resort hotels offer large discounts during the off-season period. A few electric utilities offer time-of-day and seasonal rates now. In time, this might evolve to the long-distance telephone type options now available.

Electric utilities as we know them now may be a dying breed. Nowadays traditional methods of producing electricity are experiencing heavy competition from alternative or more efficient technologies. And utilities don't offer many choices for consumers to lower their costs and obtain services more appropriate to their needs. If utilities are to meet the need outlined earlier for new marketing approaches, dramatic changes are needed. They need restructuring, and deregulation may be the best way to ensure their ability to thrive in the new energy future. Interestingly, the public power companies that are members of the APPA have, in general, shown the greatest recognition of their changing market—unlike the investor-owned utilities. Few utilities of any kind, however, have moved aggressively to make the necessary changes.

Fortunately, oil and gas companies are somewhat better structured to face the new market opportunities. These companies have traditionally sold oil products like gasoline and heating oil. Perhaps gas and oil companies also ought to redefine their products from sellers of petroleum products to providers of energy services. Will we see gas stations become renters of vehicles on a cost-per-mile basis? Could natural gas companies offer steam and electricity from gas-fired cogeneration systems? Will heating oil companies begin retrofitting houses and installing new furnaces—even those fired by natural gas or electric heat pumps—suited to whatever best served the customer? The answer is yes—and some companies are doing these things now.

Yet whenever the subject of allowing oil companies to expand into businesses other than producing, refining, or selling oil, many Americans become concerned. Because of the oil industry's huge economic resources and power, people become alarmed that they will exercise control over other fields in such a way as to limit competition, thus keeping prices higher and customer service options lower than necessary. While these are legitimate concerns, it should be remembered that the businesses they are in—selling oil and natural gas—are on the decline. These businesses are also very competitive. Atlantic Richfield (ARCO) made this very clear when they took the bold step of eliminating credit cards solely to lower their pump prices 3 to 5 cents per gallon. It was a dramatic success. ARCO has demonstrated that the market is highly competitive, that a few cents a gallon matter. These discounts boosted ARCO's gasoline sales in company-owned stations by 50 percent. Amoco, Mobil, and Gulf are all offering various experimental cash discount schemes. Exxon customers get a 4-cent-per-gallon price break with their new cash program, while credit card customers are penalized by being charged a higher fee. By their offer, Exxon is also demonstrating its conviction that the marketplace is highly competitive.

It may be best for the country if oil company resources are used in more innovative ways. No one benefits if an automobile company is not allowed to go into leasing, maintenance contracts, and insurance. Limiting the dealer and its sales-people to sales of cars will only be a waste to our economy and unprofitable to the dealer. Similarly, if oil and gas companies reach the same conclusions we do, they need to be free to shift resources into areas that will better serve the needs of their customers, including transfering some of their assets into energy services.

In order for a fairly secure energy future to emerge, attitudes must continue to change and myths must die. Yet the public's sense of direction is right and can be observed by the results we see and experience: self-interest has taken root. What we will see is more and more people participating in the sharing of its benefits. It is a process we call "creating abundance."

11

Postscript

In reviewing all we have said, we have no reservations about our claim that the energy issue will recede from importance. But the real question is how similar resource problems will be handled in the future. Have we learned something from our energy experience to help avoid making the same political mistakes all over again?

The water issue provides a good example. Underground water acquifers are now receding, and more and more wells have been going dry. In several parts of the country, crop production has fallen each year because the water being pumped has become increasingly saline. Therefore, water shortages could expand from a periodic regional concern in the 1980s to a national crisis sometime in the 1990s. Yet, in the face of all these warning signs, in most areas water is priced as though there were no limits on its supply.

If a national water crisis does happen, how will our government respond? Unfortunately, while we were upbeat throughout this book, we must now turn decidedly pessimistic. Our musings conjure up a scenario that might go something like this:

- Sometime in the 1990s, an especially severe nationwide drought will cause the media to label the shortage a "crisis."

- After a few days of trying to downplay the situation, the President will finally schedule an address to the nation and announce the creation of a cabinet-level Water Department and a trillion-dollar research program to deal with the shortage.

- Several Senate and House committees will then hold prolonged

163

hearings to get in on the act. Regionalism will rear its head as lawmakers argue for or against proposals to pipe water from the northwest to southern California or from the Mississippi River to Texas. A Massachusetts senator will point out that the 17 western states consume 84 percent of the nation's water supply.

- Californians will be asked to testify how they dealt with the drought of 1977 and whether those methods have any applicability. They will all say yes but disagree totally as to which remedies really made a difference.

- Most members of Congress, with strong media support, will declare that, "The shortage of water in this country will only get worse unless our government can come to grips with the problem and provide some solutions."

- Farm organizations will propose that a national water allocation program be established. Under the plan, agricultural users would receive 100 percent of their needs; city residents, 80 percent of their previous year's consumption; industry, 63 percent of their previous year's consumption; and no water will be available for recreational purposes. Arguments will arise as to whether growers of tobacco, flowers, and thoroughbred horses would be classified as agricultural users and thus be allocated 100 percent of their needs. Food processors will suggest that food manufacturers and distributors of these products be classified with agriculture, since it would do little good to produce food if it could not be processed.

- The executive branch of government will strongly oppose an allocation or rationing program, noting that the bureaucracy to administer such a program would cost billions of dollars a year. They will suggest instead that an expanded water research budget would make a major contribution to resolving the difficulty.

- Scientists will testify that the crisis could be solved by building a number of desalination plants on the Pacific and Atlantic coasts and shipping the water to the interior of the country by pipeline. Little attention will be paid to the cost of water services under such alternatives. Most such proposals will be justified on their technical feasibility and their creators' belief that the costs would be insignificant in relation to the cost of a nation without water. Some will suggest drilling wells to deeper aquifers as another alternative.

- Someone from a Washington think tank will suggest that there are a number of technologies available for reducing water use that could be adopted by individuals, companies, and farmers if the price of water were allowed to rise to its market level, presumably high enough to make these technologies economical to install. For instance, closed water purification systems for homes will be advocated that reduce fresh water needs of the average home to lĕss than 5 percent of that used during the 1970s and 1980s. It will be argued that numerous retrofits to current industrial processes will reduce water consumption 10 to 50 percent. A university study will show that drip irrigation methods reduce water needs dramatically without offsetting production declines.

- A suggestion to let water prices rise to market levels over a 3-year period will be rejected by most members of Congress with the argument that only the rich would be able to afford water under such a plan and the poor would suffer great hardships. They will propose federal controls on water prices, reaffirming their belief that we cannot trust this problem to be resolved by market forces. As a compromise, some legislators will suggest that we should mandate the adoption of some water-saving devices, forcing the large water departments in each city to give retrofit loans and technical assistance.

Unfortunately, this scenario is more likely to occur than not. The nation will probably not have learned sufficiently from the energy crisis of the 1970s to handle the next crisis with any more aplomb. As the foregoing suggests, we may still define the problem more narrowly than we should, thus leading to less than optimum solutions. What is more critical, we may still be afraid to follow the advice of Freud, Niebuhr, and Adam Smith or to heed our lessons from the energy crisis—that is, to allow self-interest to work in response to price.

In spite of all the evidence available to the contrary, 14 major environmental groups concluded in their joint 1982 report on President Reagan's energy policies and programs, that "the Administration and the Congress must recognize that the nation's energy future is not just a matter of chance, determined by uncontrollable market forces, or unpredictable resource shortages. . . . What is needed most is a recognition by our national leaders that the important choices before us cannot be blindly left to the market place, and that by planning wisely, we can safeguard the national interest while meeting our energy needs."[1]

Of course we disagree. The future we have sketched is one that relies

---

[1] "The Reagan Energy Plan: A Major Power Failure," a report from the Center for Renewable Resources, Cousteau Society, Environmental Action, Environmental Action Foundation, Environmental Defense Fund, Federation of American Scientists, Friends of the Earth, National Audubon Society, Natural Resources Defense Council, The Nuclear Club, Inc., Nuclear Information and Resource Service, Sierra Club, Solar Lobby, and Union of Concerned Scientists, Washington, D. C., Mar. 24, 1982, p. 38.

not on the "blind" choices of consumers but on intelligent ones, consistent with the consumer's desire to minimize costs. And this future looks remarkably bright. It is a future in which the environment is safeguarded, national security is protected, imports of oil fall, the poor are aided with dignity, government bureaucracy and decision-making is minimized, and consumer costs for energy services are lowered. If (contrary to the wishes of the 14 environmental groups) present energy policy were based on a truly free market—with no nuclear, synthetic fuel, solar or conservation subsidies—we would drop the few remaining tentative statements in this book.

If the environmentalists were to argue the same way about water, we would have a similar agreement. If markets take over after the water scenario we have sketched runs its course, consumers of water will most likely again be blessed with an increasing supply of choices, and competition among those choices will ensure that water services of optimum quality will be available at the lowest possible cost. If so, this highly disbursed process will again overcome the political mistakes of central government and lead to water sufficiency.

Index

ABOUT THE AUTHORS

ROGER W. SANT is Chairman and President of Applied Energy Services, Inc., which he founded in 1981 to sell power to industry from its cogeneration facilities. From 1977 to 1981, Mr. Sant was director of research for the Mellon Institute's Energy Productivity Center. He also served as President Ford's top official for energy conservation. He is a member of the Aspen Institute's Committee on Energy and holds an M.B.A. from the Harvard Graduate School of Business Administration.

DENNIS W. BAKKE is Executive Vice President and Chief Operating Officer of AES, Inc. In the Federal Energy Administration, he served as Deputy Assistant Administrator for Energy Conservtion and Environment. Mr. Bakke was Deputy Director of the Energy Productivity Center of the Mellon Institute. He holds degrees from the Harvard Graduate School of Business Administration and the National War College.

ROGER F. NAILL is Vice President for Energy Planning Services at AES, Inc. He has served in the U.S. Department of Energy and as Director of Dartmouth College's Energy Policy Project. He holds a Ph.D. in Engineering/Operations Research from Dartmouth College and an M.S. in Management from MIT's Sloan School of Management. Dr. Naill is the author of numerous articles and books on energy issues.

JAMES BISHOP, JR., is President of Bishop Associates, a firm concentrating on education and analysis of energy issues and media relations within the field. He was Director of Public Affairs for the U.S. Department of Energy and formerly national energy correspondent for *Newsweek*. Mr. Bishop has also written on energy issues for the *Christian Science Monitor*, the *Boston Globe*, and the *Atlanta Constitution*.